Handbook of Drugs
for
Tropical Parasitic Infections

Handbook of Drugs for Tropical Parasitic Infections

Second Edition

Yakoub Aden Abdi
Lars L. Gustafsson
Örjan Ericsson
Urban Hellgren

CRC Press
Taylor & Francis Group
Boca Raton London New York

CRC Press is an imprint of the
Taylor & Francis Group, an **informa** business
A TAYLOR & FRANCIS BOOK

First Published 1995 by Taylor & Francis

Published 2021 by CRC Press
Taylor & Francis Group
6000 Broken Sound Parkway NW, Suite 300
Boca Raton, FL 33487-2742

ISBN 13: 978-0-7484-0168-0 (pbk)
ISBN 13: 978-0-7484-0167-3 (hbk)

Visit the Taylor & Francis Web site at
http://www.taylorandfrancis.com

and the CRC Press Web site at
http://www.crcpress.com

British Library Cataloguing in Publication Data

A catalogue record for this book is available from the
British Library

Library of Congress Cataloging in Publication Data are available

Typeset by Mathematical Compositions Setters Ltd,
Salisbury, Wiltshire, SP3 4UF

Contents

Preface

The second edition of *Handbook of Drugs for Tropical Parasitic Infections* is a product from the Unit of Tropical Pharmacology at the Department of Clinical Pharmacology, Huddinge University Hospital. The unit is a collaborative venture between the Departments of Infectious Diseases and Clinical Pharmacology, and the Hospital Pharmacy. Our department has been involved for many years in research on drugs used in the treatment of tropical parasitic infections. The emphasis has been to develop and apply new bioanalytical techniques to study the clinical pharmacokinetics and metabolites of old and new drugs. Research fellows from Africa, Asia, and South America have participated in this work giving us important feedback from areas where tropical diseases are endemic. Dr Yakoub Aden Abdi from Somalia is one of these past fellows who has devoted his research on the re-evaluation of old antiparasitic drugs. It is an honour that he and his Swedish colleagues asked me to write this Preface.

During the past 40 years novel drugs have been introduced for diseases that were in the past the cause of death of thousands of people. Advances in the field of clinical pharmacology have contributed to a safer and more effective use of both old and new drugs and thereby to better patient care. In particular, new knowledge about genetic and environmental determinants of drug metabolism in humans has made it possible to introduce rational strategies in drug treatment. Pharmacoepidemiology, a science concerned with epidemiological aspects of the safety and efficacy of drug products and their utilization in the population, has also grown in importance in recent years. Developed and less developed countries seem to share a number of problems leading to irrational drug use such as old-fashioned teaching in pharmacology, drug information that is product- rather than problem-oriented and increasing criticism among patients and politcians about how drugs are being prescribed by physicians.

Modern drug therapy for tropical parasitic infections started almost 200 years ago with the isolation of quinine. Since then, more powerful drugs have been introduced. However, the rate at which new drugs have been developed for these infections has been relatively slow, and millions of people are still suffering because of parasitic infections such as malaria, schistosomiasis, trypanosomiasis and onchocerciasis. Most of the drugs that are available for such plagues are old and have complicated and empirically derived dosage regimens. Recent data on their pharmacokinetics, and re-evaluation of the use of these drugs in the field, reveal that their effectiveness can be improved and their safety increased by relatively simple measures. The second edition of this handbook aims to provide well-evaluated information about the pharmacological properties and the therapeutic use of drugs used for tropical parasitic infections. It is hoped that the book complies with the ideology of evidence-based medicine.

I would like to express my gratitude to the authors, who have devoted much of their spare time to the writing of this book.

Folke Sjöqvist, MD, PhD
Professor of Clinical Pharmacology,
Director of the WHO Collaborating Centre in Drug Utilisation Research and Clinical Pharmacological Services,
Huddinge University Hospital

Huddinge,
June 1995

Acknowledgement

The production of this book has been made possible with grants from the Swedish Agency for Research Co-operation with Developing Countries (SAREC), the National Corporation of Swedish Pharmacies (Apoteksbolaget AB) and the WHO Collaborating Centre for Clinical Pharmacological Services and Drug Utilisation at Huddinge University Hospital.

The literature search and collection of original papers were carried out by the Drug Information and Research Centre (DRIC) at the Department of Clinical Pharmacology, by Elisabeth Törnqvist. We are particularly indebted to Professor Folke Sjöqvist who encouraged us from the beginning to write this new edition and who was kind enough to write the Preface for the book. We are also indebted to Associate Professor Gunnar Alván, director of DRIC for reading the book and sharing with us his valuable comments and views. Drs Mohammed Hassan Alin, Geoffrey Edwards, Birgitta Evengård and Evert Linder have all read different parts of the book and are acknowledged for their contributions. We are also grateful to Mrs Margareta Fogelström for technical assistance in typing the manuscript at its final stages and to Ingrid Hasselberg for checking the commercial preparations of the drugs. Valuable help in drawing the chemical structures of the drugs was provided by Inger Vikström from the hospital pharmacy.

Although the second edition of *Handbook of Drugs for Tropical Parasitic Infections* is the product of contributions from many people, any errors or questionable evaluations encountered in the text or the chemical structures are the responsibility of the authors alone. We will gladly welcome any comments or advice on the contents or layout of the book from our readers.

Yakoub Aden Abdi
MD, PhD

June, 1995

Abbreviations

The following abbreviations are those that appear in several monographs. There are others which appear in single monographs and they are described when they appear for the first time in the monograph.

5-HT	5-Hydroxytryptamine
CNS	Central nervous system
CSF	Cerebrospinal fluid
DNA	Deoxyribonucleic acid
ECG	Electrocardiogram
G-6PD	Glucose-6-phosphate dehydrogenase
GABA	Gamma-aminobutyric acid
GC	Gas chromatography
GC/MS	Gas chromatography–mass spectrometry
HPLC	High-performance liquid chromatography
i.m.	Intramuscular
i.v.	Intravenous
MAO	Monoamine oxidase
MW	Molecular weight
RBC	Red blood cells
RNA	Ribonucleic acid
SDX/PYR	Sulphadoxine/Pyrimethamine
TDR	Tropical Diseases Research Unit
WHO	World Health Organization

Introduction

The aim of the new edition of *Handbook of Drugs for Tropical Parasitic Infections*, remains the same as its predecessor. It is largely designed to give physicians, pharmacists, health workers, medical students and nurses in developing countries refined and abbreviated information about drugs used for parasitic infections highly prevalent in their environment. The authors hope that the book will also be useful for clinicians or medical students in non-endemic areas who need information about drugs that is normally not included in their local therapeutic guidelines.

Development of antiparasitic drugs

Many of the drugs used for the treatment of tropical parasitic infections were introduced more than 30 years ago. Most of them are toxic and have complicated dosage regimens. Some drugs like melarsoprol, suramin, pentamidine and pentavalent antimonials have to be given parenterally for prolonged periods of time. With such treatment regimens, and the fact that most of these drugs are toxic, it is often difficult to complete the treatment. Because of the low economic incentive, pharmaceutical companies have shown little interest in developing new drugs to control diseases prevalent in less developed countries. Despite this, there has been notable progress in research in parasitic diseases and a few important drugs have been introduced for some diseases during the last two decades. This has largely been due to the efforts of the Tropical Diseases Research Unit (TDR) at the WHO in Geneva. Notable examples are the great hope raised by the recent introduction of more effective and safer drugs such as artemisinin, praziquantel, eflornithine and ivermectin. Ivermectin alone may have saved tens of thousands from blindness during the last few years. Even more exciting is the hope that a malaria vaccine may become available in the not too distant future.

Rational use of antiparasitic drugs

Although the availability of safe and effective drugs for tropical parasitic infections is limited, better understanding of the few that are presently used will enhance their efficacy and reduce their toxicity. With the development of specific analytical methods for some of the drugs, it is becoming possible to study the disposition of these drugs in relevant patients. Knowledge about the pharmacokinetics of these drugs will help us design optimal dosage regimens. In most of the developing countries, drugs are sold in several different brand names. Since the bioavailability (bioequivalence) of the different commercial preparations might vary, it is important that physicians prescribe only generic names. Many drugs are also available commercially as salts. In such a case it is important that the dosage should be calculated as the base. Unfortunately this might not be clear in most handbooks, and the physician must be aware of this problem. In rural areas, the choice of the route of drug administration is also an important factor for the success of the treatment. Intravenous administration of drugs is generally not feasible in rural areas because of shortages of trained personnel. Repeated use of syringes is also common and can be the source of spread of hepatitis or AIDS infections. Parenteral administration of drugs is expensive and may deter patients from seeking treatment. It is very important that alternative routes of drug administration should be investigated, e.g. rectal preparations especially for children.

Poor patient compliance is another major problem with drugs used for tropical parasitic infections, but the extent is unknown. Drugs with favourable treatment schedules, i.e. single dose regimens, should be preferred. A fixed dosage regimen is the norm rather than the rule for most of the drugs used for the tropical parasitic infections. It is well known that body weights of patients in developing countries are on average much less than those of people in the western world. Even in developing countries, large variations may exist between people in urban areas and those in rural places where undernutrition, malnutrition and diseases are more prevalent. Thus fixed dosage regimens for all patients do not seem rational and will definitely cause overdosing in some patients. For this reason, it is important to individualize drug therapy. Because of possible genetic reasons, it is possible that some patients might not be able to metabolize certain drugs. It is therefore important that physicians are aware of such therapeutic problems and should think of this possibility in the event of a patient with unexplained toxicity.

Sources of information

The information summarized in this book has been collected largely by the staff of the Drug Research & Information Centre (DRIC) at the Department of Clinical Pharmacology, Karolinska Institutet at Huddinge University Hospital.

The information summarized in the different monographs was retrieved from:

1. Biomedical journals:
   A renewed medline search was made covering the time from the first edition (1986). Old references which were deemed not valid or outdated have been excluded.
2. Handbooks consulted:
   Therapeutic Drugs, edited by Sir Colin Dollery (1991), (London: Churchill Livingstone).
   Martindale: The Extra Pharmacopoeia, 30th edn (1993), (London: Pharmaceutical Press).
   Goodman & Gilman's The Pharmacological Basis of Therapeutics, 8th edn, edited by A. G. Gilman, T. W. Rall, A. S. Nies and P. Taylor (1990), (New York: Pergamon Press).
   Meyler's Side Effects of Drugs, 12th edn, edited by M. N. Dukes (1992), (New York: Elsevier).
   Drugs in Pregnancy and Lactation, 3rd edn, edited by G. G. Briggs, R. K. Freeman and S. J. Yaffa (1990), (London: Williams and Wilkins).
   WHO Model Prescribing Information. Drugs used in parasitic diseases (1990), (Geneva: World Health Organization).

The plan of the book

The general layout of the book remains the same as that of its earlier edition. It starts with a chapter on drug recommendations. This chapter is intended to serve the user as a quick reminder of the main line of drugs used for each parasitic infection. The body of the book contains monographs detailing pharmacological information available for 38 drugs. The monographs are arranged in alphabetical order. Each monograph is further subdivided into seven sub-headings. Some of the drugs such as the tetracyclines (tetracycyline and doxycycline) and antimonials (sodium antimony gluconate and meglumine antimoniate) are described in the same monographs. Amodiaquine, dapsone, niridazole, hycanthone, mepacrine, tryparsamide, and trivalent antimonials have been excluded since they are no longer used and safer and more effective drugs have become available. Some new preparations such as eflornithine, amphotericin B, halofantrine, and doxycycline have been included. The monographs on ivermectin, mefloquine, and artemisinin (qinghaosu) and its derivatives have been substantially expanded. Below we describe briefly the sub-headings of the different monographs:

Chemical structure and physical properties

The structural formula is given for each compound. The molecular weight is given for the drug itself and for those salts which are used in pharmaceutical preparations. Most of the drugs are bases, only a few are acids or neutral compounds. The pKa is stated when it is known. Many drugs are sensitive to light and humidity as indicated by the brief storage recommendations. As a general rule all drugs should be protected from direct sunlight. This is especially important in a warm and humid climate. Note that some drugs for injection, although stable in the dry state, degrade rapidly after preparing a solution. In such cases the solution has to be used immediately after preparation.

Pharmacology and mechanism of action

In this section the reader finds the main pharmacological effects of each drug as shown *in vitro* or *in vivo* (animals). However, all the pharmacological activities listed may not be useful in man. Clinically, the drug should only be used for the diseases mentioned in the section 'indications'. For most drugs, the mechanism of action is still unknown. Since the publication of the previous edition very little progress has been made in this area and we still do not know very much about how most of these compounds kill parasites. There are few exceptions where mechanisms of action are known and these include antifolate drugs (proguanil, pyrimethamine and sulphadoxine), chloroquine and quinine.

Pharmacokinetics

In order to obtain reliable pharmacokinetic data it is necessary to determine drug concentrations in biological fluids with an analytical method that is specific, i.e. one that determines the drug concentration without interference by endogenous compounds, metabolites or other drugs. Usually chromatographic methods such as high-performance liquid chromatography and gas chromatography are regarded as specific. It is indicated in the monographs whether or not specific analytical methods have been developed, and reference is given to one or more methods. In those cases where it is stated that specific methods do not exist, the pharmacokinetic data, if described, must be regarded as uncertain.

Pharmacokinetic data is important in designing an optimal dosage regimen. Knowing the routes of drug elimination and excretion is also important as to avoid overdosing in patients with special problems such as kidney impairment or liver failure who may accumulate the active form of the drug in the body. Pharmacokinetic data of a drug may also explain the lack of effect or increased toxicities that may be observed in some patients. Such patients may be metabolizing or eliminating the drug differently from the rest of the population. This could be due to genetically determined differences in the metabolic capacity of those individuals, or interacting environmental factors such as nutritional status or concomitant intake of other drugs.

Clinical trials

Conducting clinical trials in rural endemic areas is generally difficult and this is one reason why most studies reported are of poor design and with limited number of patients. Moreover, most of the drugs used today for tropical parasitic infections have been introduced several decades ago when today's sophisticated ways of drug evaluations, i.e. randomized controlled studies were not available. Most of the studies are open, therefore the results must be interpreted with extra caution.

Indications

Only indications for which the drug has been shown to be effective and which have been recommended by the WHO have been included. Other indications may be listed in textbooks and in pamphlets from pharmaceutical companies. However, supporting evidence for the effectiveness of the drugs for these indications is sometimes very unsatisfactory.

Pregnancy and lactation

Teratogenicity is difficult to detect since it usually occurs at a low frequency. Animal data are a good indication of risk, but animal studies can not be directly extrapolated to humans. As a general rule, drug treatment during pregnancy should be avoided. In most cases this is not possible. Where possible we provide information and our experience of the drug during pregnancy both in animals and in humans. However, it is the responsibility of the physician to make the best judgement of the situation comparing the existence of any risk of malformation against the need for the treatment.

Side effects

Side effects are common with most antiparasitic drugs, but may be even more frequent than generally reported. Proper studies evaluating the incidence and severity of the side effects in a controlled manner are rare. In some diseases, it may be difficult to distinguish between the symptoms due to the disease and the side effects of the drug. The side effects reported in the book are those extracted from reported clinical trials and case reports.

Contraindications and precautions

Absolute contraindications are rare for most drugs. In some situations withholding the treatment might be more dangerous than any damage that the drug might cause to the patient. Proper understanding of the pharmacological actions of the drug and its disposition in humans will avert serious mistakes in dosing, i.e., overdosing in patients with kidney or liver diseases. Thus, it is important that the clinician is aware of the pharmacological properties of the drug.

Interactions

Polypharmacy is a common phenomenon in many of the developing countries where national drug policies usually do not exist or are not enforced. In such cases drug interactions can occur. Many traditional herbs used as a medicine may also interact with the drugs, but this is a poorly investigated area. Drug interactions at the metabolic level seem to be most important, especially drugs and other xenobiotics metabolized by the same cytochrome P450 isoenzymes.

Dosage

The dosage regimens in this book have in most cases been taken from the World Health Organization recommendations. However, good dose-finding studies are lacking for many drugs and the dosage schedules are too often based on clinical experience only. Where it has been considered appropriate we have also mentioned recommendations from *Martindale: The Extra Pharmacopoeia* (London: Pharmaceutical Press), Dollery and original articles. Dosage should preferably be expressed as amount of free base or acid rather than amount of salt. Unfortunately it is common practice to express dosage of some drugs as salt or hydrate. This is a source of confusion and we would like to stress that the dosage recommendations must be read with great care in order to avoid the risk of mistakes.

Preparations

Information about the commercial preparations of these drugs have been obtained from several different sources, e.g., *Martindale: The Extra Pharmacopoeia*, Dollery: *Therapeutic Drugs*, and databases at the National Corporation of Swedish Pharmacies (Apoteksbolaget AB). For some of the drugs we have made direct contacts with the manufacturers. For some drugs only one or a few preparations exist, while for some frequently used drugs like chloroquine, quinine and metronidazole, several preparations are available and it has not been possible to list them all. We assume that every physician is well aware of the preparations of these drugs which are sold locally.

Drug recommendations

The tropical parasitic infections are classified as protozoal and helminthic. For some infections several drugs might be available. The choice between them should not only depend on the efficacy and safety as special consideration must be given to the cost and the local availability of the drug. Therefore, the listed drugs are not given as first, second or third choices.

Recommended dosage schedules are given in the monograph on each drug. Consult the relevant monograph to ascertain whether the doses are expressed as a salt or as a base, since the administered dose may vary substantially between different preparations.

Protozoal infections

Infection (parasite)	*Generic drug name*	*Comments*
Amoebiasis (*Entamoeba histolytica*)		
1. Acute intestinal	Metronidazole Tinidazole	Diloxanide should be given subsequently
2. Hepatic	Metronidazole Tinidazole	
	Dehydroemetine +chloroquine	This combination should be used only as a last resort
3. Asymptomatic carriers	Diloxanide	In non-endemic areas
Giardiasis (*Giardia lamblia*)	Tinidazole Metronidazole	
Leishmaniasis		
1. Visceral (kala-azar) (*Leishmania donovani*)	Pentavalent antimonials Amphotericin Pentamidine	Amphotericin and pentamidine should be used only in cases unresponsive or intolerant to pentavalent antimonials
2. Mucocutaneous (*Leishmania braziliensis*)	Pentavalent antimonials	Mucocutaneous leishmaniasis due to *L. aethiopica* should be treated with pentamidine

Infection (parasite)	*Generic drug name*	*Comments*
3. Cutaneous (*Leishmania major*) (*Leishmania mexicana*) (*Leishmania tropica*)	Pentavalent antimonials	Cutaneous leishmaniasis due to *Leishmania aethiopica* should be treated with pentamidine. Preliminary data show also that ketoconazole* may be more effective against *Leishmania mexicana*
Malaria		
1. Treatment (*Plasmodium vivax, P. ovale* and *P. malariae*)	Chloroquine Mefloquine	Mefloquine for treatment of chloroquine resistant *P. vivax*
2. Radical cure (*P. vivax, P. ovale*)	Primaquine	After initial treatment with chloroquine
3. Treatment (*P. falciparum*)		
(a) Chloroquine sensitive areas	Chloroquine Quinine	Quinine is used in severe or complicated malaria
(b) Chloroquine resistant areas	Sulphadoxine/pyrimethamine Halofantrine Mefloquine Quinine Quinine plus SDX/PYR	Drug selection depends on the parasite susceptibility Quinine is used in severe or complicated malaria
(c) Areas with multiresistance in South East Asia	Artemisinin derivatives Halofantrine Mefloquine Quinine Quinine + tetracyclines	Quinine plus tetracyclines or artemisinin derivatives are used in severe or complicated malaria
4. Prophylaxis (*P. falciparum*)		
(a) Chloroquine sensitive areas or areas with low transmission	Chloroquine	
(b) Chloroquine resistant areas with significant transmission	Chloroquine+ proguanil Mefloquine	SDX/PYR can be used for self-treatment in some areas

Infection (parasite)	*Generic drug name*	*Comments*
(c) Multiresistant areas in South East Asia with significant transmission	Mefloquine Doxycycline	
5. Prophylaxis (*P. vivax*, *P. ovale*, *P. malariae*)	Chloroquine	Mefloquine is more effective in areas where chloroquine resistant *P. vivax* is common
Trypanosomiasis		
1. *Trypanosoma cruzi* (Chagas disease)	Nifurtimox	
2. *Trypanosoma gambiense*		
(a) Early stage (haemolymphatic)	Pentamidine Suramin	
(b) Late stage with CNS involvement	Eflornithine Suramin + melarsoprol	
3. *Trypanosoma rhodesiense*		
(a) Early stage (haemolymphatic)	Suramin Eflornithine Pentamidine	
(b) Late stage with CNS involvement	Suramin + melarsoprol Suramin + eflornithine	

* See under antimony compounds.

Helminthic infections

Infection (parasite)	*Generic drug name*	*Comments*
Nematode infections		
1. Intestinal diseases		
(a) Ascariasis* (*Ascaris lumbricoides*)	Albendazole Mebendazole Levamisole Piperazine Pyrantel	Mebendazole is preferred for the treatment of patients with polyinfections, e.g., ascariasis, hookworms, and trichuriasis
(b) Enterobiasis (*Enterobius vermicularis*)	Albendazole Mebendazole Pyrantel Piperazine	
(c) Hookworm (*Ancylostoma duodenale*)	Albendazole Mebendazole Pyrantel Bephenium	
(*Necator americanus*)	Albendazole Mebendazole Pyrantel	
Cutaneous larva migrans	Thiabendazole Albendazole	
(d) Strongyloidiasis (*Strongyloides stercoralis*)	Thiabendazole Albendazole	
(e) Trichinosis (*Trichinella spiralis*)	Thiabendazole Albendazole Mebendazole	Usually together with corticosteroids
(f) Trichuriasis (*Trichuris trichiura*)	Albendazole Mebendazole	
2. Extraintestinal diseases		
(a) Dracunculiasis (*Dracunculus medinensis*; guinea worm)	Metronidazole Niridazole	Drug treatment is of limited value
(b) Filariasis		
Lymphatic filariasis	Diethylcarbamazine	
	Ivermectin	Not effective against adult filariae
Loa loa	Diethylcarbamazine Ivermectin	

Infection (parasite)	*Generic drug name*	*Comments*
Onchocerciasis (*Onchocerca volvulus*)	Ivermectin Diethylcarbamazine	Both drugs are only effective against the microfilariae. In severe cases suramin might be used to kill the adult filariae
'Tropical eosinophilia'	Diethylcarbamazine	
Cestode infections		
1. Adult (intestinal stage)		
Diphyllobothrium latum *Hymenolepis diminuta* *Hymenolepis nana* *Taenia saginata* *Taenia solium*	Niclosamide Praziquantel	
2. Larval (tissue stage)		
Echinococcus granulosus (hydatid cyst)	Albendazole Mebendazole	
Echinococcus multilocularis	Albendazole	Of limited value. Surgical excision is the only reliable means of treatment
Taenia solium (cysticercosis)	Praziquantel Albendazole	
Trematode infections		
Fascioliasis	Triclabendazole**	Still under clinical evaluation
(*Fasciola hepatica*)	Bithionol Praziquantel	
Fasciolopsiasis (*Fasciolopsis buski*)	Praziquantel Niclosamide	
Opistorchiasis (*Opisthorchis* (= *clonorchis*) *sinensis*)	Praziquantel	
Paragonimiasis (*Paragonimus westermani*)	Praziquantel Bithionol	
Schistosomiasis		
Schistosoma haematobium	Metrifonate Praziquantel	In single individuals and in situations where re-infection is not a problem praziquantel should be used

Infection (parasite)	*Generic drug name*	*Comments*
Schistosoma intercalatum	Praziquantel	
Schistosoma japonicum	Praziquantel	
Schistosoma mansoni	Oxamniquine Praziquantel	In single patients and in situations where re-infection is not a problem praziquantel should be used
Schistosoma mekongi	Praziquantel	

* Small repeated doses are recommended for children with large worm loads, otherwise intestinal obstruction may occur.

** Still under clinical evaluation and is not discussed in the book.

Albendazole

Chemical structure

Albendazole

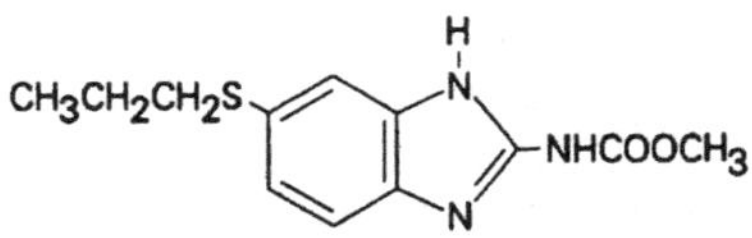

Albendazole sulphoxide (active metabolite)

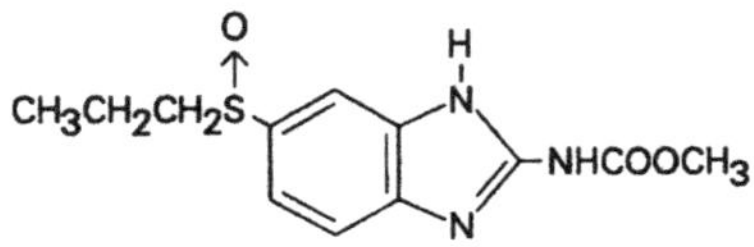

Physical properties

MW 265; pKa not known. The drug is insoluble in water.

Pharmacology and mechanism of action

Albendazole is a benzimidazole carbamate derivative which is structurally related to mebendazole. It was originally introduced as a veterinary drug in 1975 and later as a human anthelminthic drug. It has a wide spectrum of activity against intestinal nematodes (hook worm, *Ascaris lumbricoides*, *Enterobius vermicularis*, *Strongyloides stercoralis*, *Trichuris trichiura* and *Capillaria philippinensis*), systemic nematodes (*Trichinella spiralis* and cutaneous larva migrans) and cestodes (*Echinococcus granulosis*, *E. multilocularis* and *neurocysticercosis*) (1). Albendazole is active against both larval and adult stages of intestinal nematodes and ovicidal against *Ascaris lumbricoides* and *Trichuris trichiura* (1). Its main metabolite, albendazole sulphoxide, may largely be responsible for the pharmacological effects of the drug.

The mechanism of action of albendazole is similar to that of other benzimidazoles (see mebendazole).

Pharmacokinetics

Specific HPLC methods have been described for the determination of the active metabolite albendazole sulphoxide (2, 3, 4). Because of extensive first pass metabolism, albendazole itself is detected only in trace amounts or not at all in plasma.

After oral administration of a single dose of 400 mg of albendazole to healthy volunteers, peak plasma concentrations between 0.04 and 0.55 μg/ml of the sulphoxide metabolite were obtained after 1 to 4 hours (5). When the drug was given with a fatty meal, 2–4-fold increase in plasma concentrations were observed (5, 6). Large intra- and inter-individual variability in the plasma concentrations of albendazole sulphoxide has been reported (5, 7), and is likely to be due to its erratic absorption and possible differences in metabolic rate. Albendazole sulphoxide binds to plasma proteins up to 70% (5). During long term treatment against hydatid disease, the concentrations of albendazole sulphoxide in cyst fluid may reach levels around 20% of that in plasma (8).

Albendazole is quickly and completely oxidized to the active metabolite albendazole sulphoxide, which is further oxidized to the inactive compound albendazole sulphone. Albendazole sulphoxide is eliminated with a plasma elimination half-life of around 9 hours. The sulphoxide metabolite is excreted through the kidneys along with the sulphone and other minor metabolites. Insignificant amounts of the main metabolite may be eliminated through the bile (5). Albendazole is a partial inhibitor of microsomal enzymes, but the drug induces also the metabolism of its sulphoxide metabolite during long term treatment in hydatid diseases (9). Albendazole sulphoxide crosses the blood–brain barrier and attains a CSF concentration one-third of that in plasma (10).

Clinical trials

In an open trial, a single dose of albendazole (400 mg as tablets or suspension) was given to 1455 patients with mixed infections (11). Using the Kato–katz technique (a quantitative test) the drug was curative in enterobiasis (100%), ascariasis (92%), ancylostomiasis caused by *Necator americanus* (90%), and in trichuriasis (70%). The drug did not produce any significant adverse effects or modifications of the haematology or clinical blood chemistry. Only 6% of the patients reported side effects (11). In a multicentre, double-blind study (12), 392 children and adults from France and West Africa with single or mixed infections were treated either with a single dose of 400 mg albendazole or placebo. Cure rates after treatments were 96% for ascariasis, 96% for ancylostomiasis, 90% for necatoriasis, and 76% for trichuriasis. About 48% of the patients were infected with *Strongyloides stercoralis* and were also cured following administration of a single dose of albendazole 400 mg daily for 3 days. Children who received half the adult dose had lower cure rates. The drug did not produce any significant side effects. Similar efficacy against strongyloidiasis has also been reported in a study with a small number of patients (13).

In randomized comparative clinical studies in patients with neurocysticercosis, single daily doses between 15 and 20 mg/kg of albendazole given for 21 to 30 days (n = 36) were compared to praziquantel given as single daily doses of 50 mg/kg for 15 to 21 days (n = 37). Evaluations made 3 to 6 months later found that albendazole was significantly more effective than praziquantel in reducing the total number of cysts and resolving the symptoms (14, 15).

Single cases of patients with cutaneous larva migrans successfully treated with albendazole have been reported (16, 17, 18). Studies with proper designs and sufficient numbers of patients are needed to confirm these reports.

There is evidence that albendazole is effective against hydatid disease. The progression of the disease is arrested with considerable clinical improvement and cyst reduction or disappearance with a longer survival time, twice that of untreated patients (19). Horton *et al.* (20) have recently reviewed the treatment outcome of 253 patients with active

Echinococcus granulosus who were treated mostly with 800 mg of albendazole daily in cycles of 28 days with 14 days rest period between cycles, with a mean duration of 2.5 cycles (range 1–12). Of these, 29% were regarded as cured, 51% improved, 18% unchanged, and 2% worsened (20). In open comparative clinical trials, albendazole has been shown to be more effective than mebendazole in curing as well as in improving the general condition in such patients (21–26).

Indications

Single or mixed infections caused by *Ascaris lumbricoides*, *Enterobius vermicularis*, *Ancylostoma duodenale*, *Trichuris trichiura*. Albendazole may be effective against cutaneous larva migrans and *Strongyloides stercoralis*, but controlled studies are needed to confirm its advantage over thiabendazole. Limited data indicate that albendazole is useful in neurocysticercosis (14, 15). Albendazole seems to be the drug of choice for the treatment of inoperable hydatid cases, but its long term benefit needs further assessment.

Pregnancy and lactation

Teratogenicity and embryotoxicity has been reported in rats and rabbits (27). There have been no reports in humans. Because of its teratogenicity in animals and lack of documentation in man, albendazole should not be given during pregnancy.

Its excretion into breast milk is unknown.

Side effects

After a single dose treatment of albendazole 400 mg, minor and transient side effects such as epigastric pain and diarrhoea were seen. Less than 6% of treated patients experience these effects (11). During the treatment of hydatid disease, where higher doses are used for longer time periods, side effects were more common and severe. In two randomized double-blind multicentre phase I and II studies (21, 26) involving 139 patients given high doses of the drug, about 20% of the patients showed side effects. These included elevation of serum transaminases (6 patients), leucopenia (3 patients), gastrointestinal symptoms (8 patients), severe headache (4 patients), loss of hair (3 patients), urticaria and itching (2 patients), fever and fatigue (1 patient), and thrombocytopenia (1 patient).

Contraindications and precautions

There are no known contraindications to the drug during single dose treatment of intestinal nematodes. During treatment against hydatid disease, liver transaminases, leukocyte and platelet counts must be monitored regularly.

Drug interactions

The concomitant administration of dexamethasone has been reported to increase the plasma levels of albendazole sulphoxide by about 50%. The parent drug, albendazole which is only detected in trace amounts at normal doses has also reached measurable levels after dexamethasone administration (28).

Dosage

Ascariasis, enterobiasis, ancylostomiasis and cutaneous larva migrans

Adults and children
A single dose of 400 mg. Re-infection is common with enterobiasis; a further dose may be required after 2 to 4 weeks.

Trichuriasis

Adults and children
A single dose of 400 mg is usually sufficient. For heavier infections the treatment can be continued for 3 days.

Strongyloidiasis

Adults and children (>2 years)
A single dose of 400 mg daily for 3 days.

Hydatid disease

Adults and children
Four 28-day courses of 10–15 mg/kg daily in three divided doses separated by 14 days rest periods. The treatment duration, however, is governed by the disease and patient tolerance.

Neurocysticercosis

Adults and children
15 mg/kg daily in three divided doses for 28 days.

Preparations

- Zentel®(SmithKline Beecham). Tablets 400 mg. Suspension 2%.
- Eskazole® (SmithKline Beecham). Tablets 400 mg.

References

1. Rossignol JF, Mausonneuve H (1984). Albendazole: a new concept in the control of intestinal helminthiasis. *Gastroenterol Clin Biol*, **8**, 569–576.
2. Hoaksey PE, Awadazi K, Ward SA, Coventry PA, Orme ML'E, Edwards G (1991). Rapid and sensitive method for the determination of albendazole and albendazole sulphoxide in biological fluids. *J Chromatogr*, **566**, 244–249.
3. Hurtado M, Medina MT, Sotelo J, Jung H (1989). Sensitive high-performance liquid chromatographic assay for albendazole and its main metabolite albendazole sulphoxide in plasma and cerebrospinal fluid. *J Chromatogr*, **494**, 403–407.
4. Zeugin T, Zysset T, Cotting J (1990). Therapeutic monitoring of albendazole: A high-performance liquid chromatography method for determination of its active metabolite albendazole sulphoxide. *Therap Drug Monit*, **12**, 187–190.
5. Marriner SE, Morris DL, Dickson B, Bogan JA (1986). Pharmacokinetics of albendazole in man. *Eur J Clin Pharmacol*, **30**, 705–708.
6. Lange H, Eggers R, Bircher J (1988). Increased systemic availability of albendazole when taken with a fatty meal. *Eur J Clin Pharmacol*, **34**, 315–317.
7. Jung H, Hurtado M, Sanchez M, Medina MT, Sotelo J (1992). Clinical pharmacokinetics of albendazole in patients with brain cysticercosis. *J Clin Pharmacol*, **32**, 28–31.

8. Morris DL, Chinnery MJ, Georgiou G, Golematis B (1987). Penetration of albendazole sulphoxide into hydatid cysts. *Gut*, **28**, 75–80.
9. Steiger U, Cotting J, Reichen J (1990). Albendazole treatment of echinococcosis in humans: effects on microsomal metabolism and drug tolerance. *Clin Pharmacol Ther*, **47**, 347–353.
10. Jung H, Hurtado M, Sanchez M, Medina MT, Sotelo J (1990). Plasma and CSF levels of albendazole and praziquantel in patients with neurocysticercosis. *Clin Neuropharmacol*, **13**, 559–564.
11 Coulaud JP, Rossignol JF (1984). Albendazole: a new single dose anthelminthic. *Acta Tropica* (Basel), **41**, 87–90.
12. Pene P, Mojon M, Garin JP, Coulaud JP, Rossignol JF (1982). Albendazole: a new broad spectrum anthelminthic. Double-blind multicenter clinical trial. *Am J Trop Med Hyg*, **31**, 263–266.
13. Chanthavanich P, Nontasut P, Prarinyanuparp V, Sa-Nguank S (1989). Repeated doses of albendazole against strongyloidiasis in Thai children. *Southeast Asian J Trop Med Pub Health*, **20**, 221–226.
14. Cruz M, Cruz I, Horton J (1991). Albendazole versus praziquantel in the treatment of cerebral cysticercosis: Clinical evaluation. *Trans R Soc Trop Med Hyg*, **85**, 244–247.
15. Takayanagui OM, Jardim E (1992). Therapy of neurocysticercosis. Comparison between albendazole and praziquantel. *Arch Neurol*, **49**, 290–294.
16. Jones SK, Reynolds NJ, Olikwiecki S, Harman RRM (1990). Oral albendazole for the treatment of cutaneous larva migrans. *Br J Dermatol*, **122**, 99–101.
17. Williams HC, Monk B (1989). Creeping eruption stopped in its tracks by albendazole. *Clin Exp Dermatol*, **14**, 355–356.
18. Orihuela AR, Torres JR (1990). Single dose of albendazole in the treatment of cutaneous larva migrans. *Arch Dermatol*, **126**, 398–399.
19. Wilson JF, Rausch RL, McMahon, Schantz PM (1992). Parasitological effect of chemotherapy in alveolar hydatid disease: review of experience with mebendazole and albendazole in Alaskan eskimos. *Clin Infect Diseases*, **15**, 234–249.
20. Horton RJ (1989). Chemotherapy of echinococcosis infection in man with albendazole. *Trans R Soc Trop Med Hyg*, **83**, 97–102.
21. Davies A, Dixon H, Pawlowski ZS (1989). Multicentre clinical trials of benzimidazole carbamates in human cystic echinococcosis (phase 2). *Bull World Health Organ*, **67**, 503–508.
22. Ellis M, von Sinner W, Al-hokail A, Siek J (1992). A clinical-radiological evaluation of benzimidazoles in the management of echinococcosis granulosis cysts. *Scand J Infect Dis*, **24**, 1–13.
23. Todorov T, Mechkov G, Vutova K, Georgiev P, Lazarova I, Tonchev Z, Nedelkov G (1992). Factors influencing the response to chemotherapy in human cystic echinococcosis. *Bull World Health Organ*, **70**, 347–358.
24. Todorov T, Vutova K, Mechkov G, Tonchev Z, Georgiev P, Lazarova I (1992). Experience in the chemotherapy of severe inoperable echinococcosis in man. *Infection*, **20**, 23–24.
25. Todorov T, Vutova K, Mechkov G, Georgiev P, Petkov D, Tonchev Z, Nedelkov G (1992). Chemotherapy of human cystic echinococcosis: comparative efficacy of mebendazole and albendazole. *Ann Trop Med Parasitol*, **86**, 59–66.
26. Davis A, Pawlski ZS, Dixon H (1986). Multicentre clinical trials of benzimidazole-carbamates in human echinococcosis. *Bull WHO*, **64**, 383–388.
27. Albendazole, in *Therapeutic drugs*, edited by Sir Colin Dollery (1991), (London: Churchill Livingstone), pp. A31–A34.
28. Jung H, Hurtado M, Medina MT, Sanchez M, Sotelo J (1990). Dexamethasone increases plasma levels of albendazole. *J Neurol*, **237**, 279–280.

Amphotericin B

Chemical structure

Amphotericin B

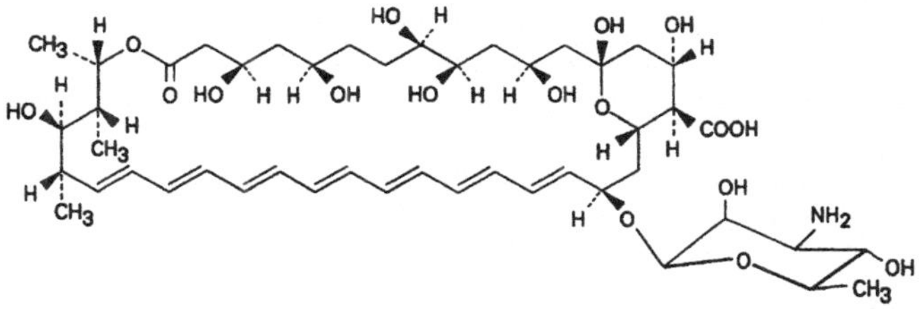

Physical properties

MW 924; pKa 5.5, 10.0. Practically insoluble in water. Store in a dark refrigerator in air-tight containers. Amphotericin B precipitates with the addition of an electrolyte solution. Precipitation has also been reported with several drugs commonly used in the tropics such as penicillin G, kanamycin, lignocaine, nitrofurantoin, oxytetracycline, and streptomycin (1). Amphotericin solutions should be used immediately after preparation.

Pharmacology and mechanism of action

Amphotericin B is a polyene macrolide antibiotic which was introduced into clinical medicine in 1955. It is primarily used for the treatment of serious systemic fungal infections. It is also used as an alternative drug for the treatment of drug resistant *Leishmania*. Amphotericin B is an effective drug, but its use is limited because of its toxicity. The advent of liposome encapsulated amphotericin may increase its use in multiresistant *Leishmania* in the future (2).

The mechanism of action of amphotericin is as yet not clear. In mycosis it binds to ergosterol present in fungal cell membranes. As a result, the drug forms pores or channels on the cell membrane which disturbs the membrane function allowing electrolytes (particularly potassium) and small molecules to leak from the cell resulting in cell death (3). Oxidative damage to the cell may also be involved in this process (4). Its mechanism of action in leishmaniasis may be similar to that in fungi.

Pharmacokinetics

A specific HPLC method has been described (5).

Because of poor oral absorption (less than 10%) and damage to the tissue after intramuscular injection, intravenous infusion is the only way for systemic administration (6).

There have been no pharmacokinetic studies in patients with leishmaniasis. The pharmacokinetic data available have largely been derived from patients with terminal cancer suffering from systemic fungal infections. After intravenous administration, the drug is distributed with an apparent volume of distribution of around 4 l/kg (7). About 90 to 95% of the drug is bound to plasma proteins, mainly to lipoproteins (8). Its access to the CSF is limited and concentrations vary between 2 and 4% of the concentration in plasma (9). The elimination is biphasic, characterized by an initial phase with an elimination half-life between 24 and 48 hours, followed by a slower phase with a half-life of up to 15 days (7). The long terminal elimination phase of the drug reflects a strong binding of the drug to body tissues. In an autopsy study, high concentrations of the drug were found in the lungs, spleen, and kidneys (10). The metabolism of the drug is as yet unknown. It is slowly excreted with the urine and the bile over a long period. Around 3% of the dose has been recovered from the urine during the first 24 hours after drug administration (7).

The drug crosses the placental barrier (11). Haemodialysis is ineffective in removing the drug from the body (12).

Clinical trials

In a prospective randomized trial in India (13), amphotericin B (14 doses of 0.5 mg/kg given i.v. on alternate days) was compared to pentamidine isethionate (20 doses of 4 mg/kg given i.m. on alternate days) in 120 uncomplicated and parasitologically confirmed cases of antimony-unresponsive visceral leishmaniasis (kala-azar). After 6 months follow-up, 46 (77%) patients treated with pentamidine were cured versus 59 (98%) patients treated with amphotericin. Amphotericin B also brought quicker abatement of fever and more complete spleen regression.

To reduce toxicity and increase its concentration in the parasite, a lipid-complexed amphotericin B has been developed recently and preliminary results are encouraging. In single individual case reports (14, 15), patients with multi-resistant visceral leishmaniasis were treated successfully and with minimal or no side effects. The patients were treated with a dose of 50 mg per day intravenously for 21 days. In a multicentre study (16), 31 patients with visceral leishmaniasis received liposomal amphotericin B. Ten patients received 1–1.38 mg/kg/day for 21 days, and another 10 received 3 mg/kg/day for 10 days. All were cured without significant adverse events and without relapse during 12–24 months of follow-up. The remaining 11 patients (immunocompromised) received 1.38–1.85 mg/kg/day for 21 days. All were initially cured, but 8 relapsed after 3 to 22 months. All patients tolerated the drug.

Indications

Treatment of visceral and mucocutaneous leishmaniasis unresponsive to standard drugs (pentavalent antimonials and pentamidine).

Pregnancy and lactation

Teratogenicity of amphotericin B in animals or in humans is unknown. Because of its toxicity, the drug should only be used if the condition of the patient makes it necessary for its use.

Its excretion into breast milk is unknown.

Side effects

Amphotericin B is highly toxic and most patients treated with the drug may experience side effects. Thus its clinical use in leishmaniasis is limited. The reported side effects are largely from patients with fungal infections. After intravenous administration, a series of adverse reactions occur. The most common ones include fever and chills, which begin an hour or two after start of infusion. Nausea, vomiting, gastrointestinal cramps, dyspnoea, bronchospasm or a true anaphylactic reaction may follow in some patients (1, 17). Nephrotoxicity is also a common side effect with rises in azotemia and decrease of about 40% of glomerular filtration rate (1). Urinary loss of potassium and magnesium may lead to severe hypokalemia and hypomagnaesemia with possible seizures. Anaemia is another common side effect which could be due to a direct suppressive effect on the erythropoietin production (1).

Most of the above side effects can be expected during treatment of patients with leishmaniasis. However, a liposome encapsulated amphotericin B seems to be effective and less toxic than conventional amphotericin B, but data are still preliminary (14, 15, 16).

Contraindications and precautions

Amphotericin B should be administered under close medical supervision. Blood urea nitrogen (BUN), haemoglobin and potassium values should be regularly monitored. During treatment with amphotericin, other nephrotoxic and potassium depleting agents should be avoided. Because of the wide range of incompatibilities reported with amphotericin B (see below), it is generally advisable not to mix it with any other drug.

Interactions

There have been no reports of drug interactions during the treatment of leishmaniasis. However, incompatibilities will occur in the infusion fluids if mixed with other substances (see physical properties).

Dosage (18)

Infusion fluids must be freshly prepared by dissolving 50 mg amphotericin B in 10 ml of sterile water and making up to 500 ml with 5% glucose to give a final concentration of 100 μg/ml solution. For adults, a starting dose of 5–10 mg is incremented by 5–10 mg daily to a maximum of 0.5–1 mg/kg. This is then infused (6–8h) on alternate days to a total of 1–3 g. (Caution: do not mix amphotericin with saline solutions, i.e. sodium chloride 0.9%, as precipitate will form).

Some centres infuse a test dose of 1 mg of amphotericin B over periods of 20 minutes to 4 hours before starting treatment. In case of intolerable toxicity with conventional amphotericin B, liposomal amphotericin B can be given by intravenous infusion (over 30 to 60 minutes) at a dosage of 1 mg/kg/day initially, increased gradually to 3 mg/kg/day for up to 21 days (1).

Preparations

- Fungizone® (Squibb). Vials containing 50 mg of amphotericin B.
- Ambisome® (Vestar). Vials containing 50 mg liposomal amphotericin B.

References

1. Antifungal drugs, in *Martindale: The Extra Pharmacopoeia*, 30th edn (1993), (London: Pharmaceutical Press), pp. 315–319.
2. Gradoni L, Davidson RN, Orsini S, Betto P, Giambenedetti M (1993). Activity of liposomal amphotericin B (AmBisome) against *Leishmania infantum* and tissue distribution in mice. *J Drug Target*, **1**, 311–316.
3. Kerridge D (1986). Mode of action of clinically important antifungal drugs. *Adv Microbiol Phys*, **27**, 1–27.
4. Brajtburg J, Powderly WG, Kobayashi GS, Medoff G (1990). Amphotericin: current understanding of its mechanism of action. *Antimicrob Agents Chemother*, **34**, 183–188.
5. Nilsson-Ehle I, Yoshikawa TT, Edwards JE, Schotz MC, Couze LB (1977). Quantitation of amphotericin B with use of high pressure liquid chromatography. *J Infect Dis*, **135**, 414–422.
6. Gallis HA, Drew RH, Pickard WW (1990). Amphotericin B: 30 years of clinical experience. *Rev Infect Dis*, **12**, 308–329.
7. Atkinson AJ Jr, Bennet JE (1978). Amphotericin B pharmacokinetics in humans. *Antimicrob Agents Chemother*, **13**, 271–276.
8. Polak A (1979). Pharmacokinetics of amphotericin B and flucytosine. *Postgr Med J*, **55**, 667–670.
9. Atkinson AJ Jr, Bindschadler DD (1969). Pharmacokinetics of intrathecally administered amphotericin B. *Amer Rev Respir Dis*, **99**, 917–924.
10. Christiansen KJ, Bernard EM, Gold JWM, Armstrong D (1985). Distribution and activity of amphotericin B in humans. *J Infect Dis*, **152**, 1037–1043.
11. Ismail MA, Lerner SA (1982). Disseminated blastomycosis in a pregnant women. *Am Rev Respir Dis*, **126**, 350–353.
12. Block ER, Bennet JE, Livoti LG, Klein WJ Jr, MacGregor RR, Henderson L (1974). Flucytosine and amphotericin B: Haemodialysis effects on plasma concentration and clearance. *Ann Intern Med*, **8**, 613–617.
13. Mishara M, Biswas UK, Jha DN, Khan AB (1992). Amphotericin versus pentamidine in antimony-unresponsive kala-azar. *Lancet*, **340**, 1256–1257.
14. Croft SL, Davidson RN, Thornton EA (1991). Liposomal amphotericin B in the treatment of visceral leishmaniasis. *J Antimicrob Chemother*, **28**, 111–118.
15. Davidson RN, Croft SL, Scott A, Maini M, Moody AH, Bryceson AD (1991). Liposomal amphotericin B in drug-resistant visceral leishmaniasis. *Lancet*, **337**, 1061–1062.
16. Davidson RN, Di Martino L, Gradoni L, Giacchino R, Russo R, Gaeta GB *et al.* (1994). Liposomal amphotericin B (AmBisome) in Mediterranean visceral leishmaniasis: a multi-centre trial. *Q J Med*, **87**, 75–81.
17 Bennet JE (1990). Antimicrobial agents. In: *Goodman & Gilman's The Pharmacological Basis of Therapeutics*, 8th edn edited by AG Gilman, TW Rall, AS Nies and P Taylor, (New York: Pergamon Press), pp. 1165–1168.
18. *WHO Model Prescribing Information. Drugs used in parasitic diseases* (1990), (Geneva: World Health Organization).

Antimony compounds

Antimonial compounds are classified as trivalent and pentavalent compounds. Examples of trivalent antimonials include potassium antimony tartrate and sodium antimony dimercaptosuccinate. These compounds have been abandoned because of their toxicity and difficulty of administration and they are not considered here. For comparative reasons, the structure of antimony tartrate is given below. Two pentavalent antimonials, sodium antimony gluconate and meglumine antimonate are commonly used. Given in equimolar doses in terms of antimony (Sb), these two compounds show similar pharmacological, pharmacokinetic and therapeutic properties. Meglumine antimonate (Glucantime) is preferentially used in French speaking countries and South America, whereas sodium antimony gluconate (Pentostam) is used elsewhere. However, the choice is only determined by their availability. Reports of one drug are applicable to the other if not otherwise specified.

Chemical structure

Trivalent antimonials

Antimony tartrate (potassium salt)

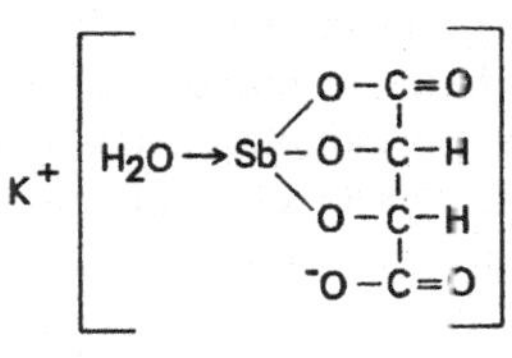

Pentavalent antimonials

Sodium antimony gluconate

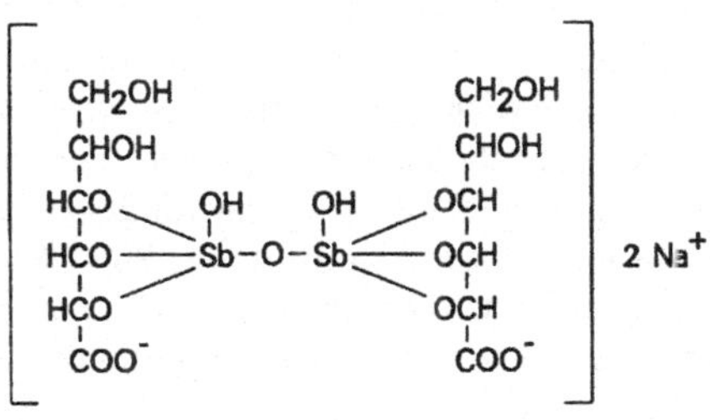

Meglumine antimonate

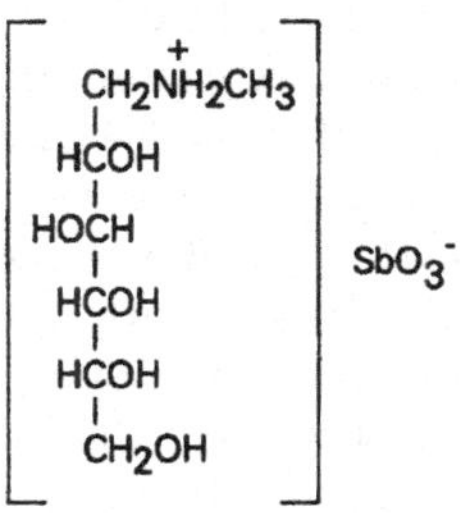

Physical properties

Meglumine antimonate: MW 366 (33% Sb). 1 g dissolves in 3 ml of water. The composition of the salt of sodium antimony gluconate is variable and thus its exact MW can not be determined. It contains 30–34% Sb and is freely soluble in water. The solutions for injection should be stored in air-tight containers and be protected from light.

Pharmacology and mechanism of action

Pentavalent antimonials are effective against *Leishmania (L) tropica* and *L. mexicana* (cutaneous leishmaniasis), *L. braziliensis* (mucocutaneous leishmaniasis) and *L. donovani* (Kala-azar or visceral leishmaniasis).

The mechanism of action of pentavalent antimonials is not fully known. These compounds interfere with the energy production of *Leishmania* amastigotes. Antimony inhibits parasite glycolytic and fatty acid oxidation activity, which leads to a decreased antioxidant defence mechanism and decreased energy for metabolism (1).

Liposome-encapsulated antimonials have been used successfully to treat *Leishmania* infections in dogs. In this form, the drug selectively concentrates in the lysosomes of the macrophages, where the parasites reside (2).

Pharmacokinetics

A specific analytical method has not been reported and the pharmacokinetic data described are based on unspecific measurements of total antimony.

Because of slow oral absorption and marked irritation to the gastro-intestinal mucosa, pentavalent antimonials are administered intravenously or intramuscularly. The pharmacokinetics of meglumine antimonate and sodium antimony gluconate are similar. Following an intramuscular injection, peak plasma levels are reached within 2 hours (3). The drugs distribute throughout the extracellular body space with a volume of distribution of 0.22 l/kg (3). Pentavalent antimonials are probably not metabolized in the body. Elimination is characterized by two phases: an initial phase with a plasma elimination half-life of around 2 hours, followed by a slow elimination phase with a half-life of between 33 and 76 hours (3, 4, 5). More than 80% of the pentavalent antimony is excreted with the urine within the first 6 hours (6). Only small amounts are excreted with the faeces (5).

Clinical trials

Visceral leishmaniasis

In a randomized clinical trial conducted in Kenya (7), 33 children and 10 adults with visceral leishmaniasis were given either 10 mg Sb/kg/day or 20 mg Sb/kg/day of sodium antimony gluconate. After about 4 weeks of treatment, 60% of those given the lower dose were cured in comparison to 75–100% of those who received the higher dose. In a study carried out in India (8), patients who received higher doses of 20 mg Sb/kg/day for 20–40 days had a cure rate of 80–97%, while the efficacy was much lower with 10–15 mg Sb/kg for a similar duration. In another study (9) by the same authors, 312 Indians with visceral leishmaniasis were divided into three treatment groups and were given sodium antimony gluconate 20 mg Sb/kg for 20, 30, and 40 days respectively. The cure rates were 87%, 94% and 98%, respectively.

Cutaneous leishmaniasis:

In a randomized, double-blind clinical trial in Panama in patients with *L. braziliensis panamensis* (10), all 19 patients treated with 20 mg Sb/kg for 20 days were cured compared to only 15 out of 21 patients treated with 10 mg Sb/kg/day for a similar treatment period. In an open study conducted in Panama (11), 51 patients suffering from leishmaniasis *b. panamensis* were treated with intramuscular sodium antimony gluconate (20 mg Sb/kg/day with a maximum dose of 850 mg Sb/day for 20 days, $n = 19$), ketoconazole (600 mg/day for 28 days orally, $n = 22$), or placebo ($n = 11$). After a 12 month follow-up, patients given sodium antimony gluconate had a cure rate of 68%, which was superior to those given placebo (0% cure rate), but inferior to those given ketoconazole (76% cure rate). Side effects were also more common in those who received the antimony preparation (11). In a randomized placebo-controlled trial, Guatemalan patients were given either sodium stibogluconate (20 mg Sb/kg/day i.v. for 20 days, $n = 32$), Ketoconazole (600 mg/kg orally for 28 days, $n = 32$), or placebo (31). The patients were followed-up for up to 52 weeks. Treatment outcome was influenced by species. Among patients infected with *L. braziliensis*, 24 of 25 in the stibogluconate group but only 7 of 23 in the ketoconazole group responded. Among patients infected with *L. mexicana*, only 4 of 7 in the stibogluconate group but 8 of 9 in the ketoconazole group responded. The number of patients included in the study was small and the effect of the drugs against *L. mexicana* was not statistically significant. Side effects were mild or moderate but were more common with those who were treated with sodium stibogluconate (12).

Mucosal leishmaniasis

In an open study (13) conducted in Panama intravenous sodium antimony gluconate 20 mg Sb/kg/day for 28 days were given to 16 patients with mild cutaneous leishmaniasis. All the patients who completed the treatment were cured. However, after a 12 month follow-up, 3 relapsed (77% cure rate). In Peru (14) 29 patients with mucous leishmaniasis were treated with similar dosages as above. Eight suffered from a mild disease of the nasal mucosa, and 21 suffered from a more severe type of the disease. After treatment only 10% of those with the severe type were cured compared to 75% of those with the mild type of the disease.

Indications

For the treatment of visceral, cutaneous and mucosal leishmaniasis.

Pregnancy and lactation

Teratogenicity has not been reported in rats (15). No malformations were reported in a child born to a mother given meglumine antimonate during pregnancy (16). Pentavalent antimonials should not be withheld from patients suffering from visceral leishmaniasis.

Small amounts of sodium antimony gluconate have been reported to be excreted in breast milk (17). Because of the poor absorption of the drug from the gut and the insignificant amounts reaching the breast milk, nursing can however be regarded safe, particularly in areas where the possibility of bottle feeding is not feasible (17).

Side effects

Pentavalent antimonials are safer than the trivalent forms. In one study (12) where the incidence of the side effects was carefully monitored, 21 out of 40 patients treated with sodium antimony gluconate complained of adverse reactions. The symptoms and signs included: phlebitis (25%), arthralgia (15%), nausea (13%), anorexia (10%), headache (8%), and rash (3%). More than half of the patients had also shown asymptomatic elevations of alanine and aspartate aminotransferases. At one point during therapy, ECG changes of T-wave flattening or inversion and prolongation of the Q–T interval were noted in more than half of the patients, but returned to normal after completion of therapy. At dosages above 20 mg/kg, the risk of cardiotoxicity increases substantially (18). Single case reports of nephrotoxicity (19, 20) and pancreatitis (21) have also been reported. Similar side effects can also be anticipated from the administration of meglumine antimonate.

Contraindications and precautions

The drug should not be given to patients with kidney failure or with cardiomyopathy. Available data suggest that dosage reductions should be proportional to the reduction in glomerular filtration rate. Slow intravenous injections (over 5–10 minutes) are necessary to avoid acute reactions such as nausea, vomiting, or substernal pain.

Interactions

Synergistic actions of pentavalent antimonials and allopurinol have been reported both in experimental *Leishmania* (22) and clinically (23).

Dosage (24)

Visceral leishmaniasis (Kala-azar)

Adults and children
20 mg Sb/kg daily (preferably in two divided doses) i.m. or i.v. (to a maximum of 850 mg) for a minimum of 20 days. Patients who relapse should be re-treated immediately with the same dose.

Cutaneous leishmaniasis (except L. braziliensis and L. aethiopica)

Adults and children
Local therapy – injection of 1–3 ml (containing 100 to 300 Sb) into the base of the lesion, repeated once, or twice if no response is apparent, at intervals of 1 to 2 days.

Systemic therapy – 10–20 mg Sb/kg i.m. or i.v. daily until a few days after a clinical cure and skin smears are negative.

Cutaneous leishmaniasis (L. braziliensis)

Adults and children:
20 mg Sb/kg daily i.m. or i.v. until the lesion is healed for at least 4 weeks. Should a relapse occur, pentamidine should be used instead.

Mucocutaneous leishmaniasis (L. braziliensis)

Adults and children
20 mg Sb/kg daily i.m. until slit-skin smears are negative and for at least 4 weeks. In the advent of toxicity or inadequate response, 10–15 mg Sb/kg should be administered every 12 hours for the same period. Patients who relapse should be re-treated for at least twice as long. Those who are unresponsive should receive amphotericin B or pentamidine.

Diffuse cutaneous leishmaniasis (L. amazonensis)

Adults and children
20 mg Sb/kg daily i.m. for several months until clinical improvement occurs.

Recently, Herwaldt *et al.* (18) have critically evaluated the different dosage regimens used by a large number of published clinical trials of pentavalent antimonials in leishmaniasis, and they concluded that the 850 mg restriction recommended by the WHO (see Dosage) should be removed. On the basis of recent efficacy and toxicological data, 20 mg Sb/kg day of pentavalent antimony given 20 days for cutaneous and visceral leishmaniasis and 28 days for mucosal leishmaniasis is recommended.

Preparations

Available as sodium antimony gluconate: 330 mg salt is equivalent to 100 mg of antimony.

- Pentostam® (Wellcome, UK). Solution for injection, 330 mg sodium antimony gluconate/ml.

Available as meglumine antimonate: 300 mg salt is equivalent to 100 mg of antimony.

- Glucantime® (Rhône-Poulenc Rorer). Solution for injection 300 mg meglumine antimony/ml.

References

1. Berman JD (1988). Chemotherapy for leishmaniasis: biochemical mechanisms, clinical efficacy and future strategies. *Rev Infect Dis*, **10**, 560–586.
2. Chapman WL, Hanson WL, Alving CR, Hendricks LD (1984). Antileishmanial activity of liposome-encapsulated meglumine antimonate in the dog. *Am J Vet Res*, **45**, 1028–1030.
3. Chulay JD, Fleckenstein L, Smith DH (1988). Pharmacokinetics of antimony during treatment with sodium stibogluconate or meglumine antimonate. *Trans R Soc Trop Med Hyg*, **82**, 69–72.
4. Goodwin LG, Page JE (1943). A study of the excretion of organic antimonials using a polarographic procedure. *Biochem J*, **37**, 198–209.
5. Otto GF, Maren TH, Brown HW (1947). Blood levels and excretion rates of antimony in persons receiving trivalent and pentavalent antimonials. *Am J Hyg*, **46**, 193–211.
6. Rees PH, Kager PA, Keating MI, Hocmeyer WT (1980). Renal clearance of pentavalent antimony (sodium stibogluconate) *Lancet*, **ii**, 226–229.
7. Manson-Bahr PEC (1959). East African Kala-azar with special reference to the pathology prophylaxis and treatment. *Trans R Soc Trop Med Hyg*, **53**, 123–136.
8 Thakur CP, Kumar P, Mishra BN, Pandey AK (1988). Rationalisation of regimens of treatment of Kala-azar with sodium stibogluconate in India: a randomised study. *BMJ*, **296**, 1557–1561.
9. Thakur CP, Kumar P, Pandey AK (1991). Evaluation of efficacy of longer duration of therapy of fresh cases of Kala-azar with sodium stibogluconate. *Indian J Med Res*, **93**, 103–110.
10. Ballou WR, McClain JB, Gordon DM, Shanks GD, Andujar J, Berman JD, Chulay JD (1987). Safety and efficacy of high-dose sodium stibogluconate therapy of American cutaneous leishmaniasis. *Lancet*; **ii**, 13–16.
11. Saenz RE, Paz H, Berman JD (1990). Efficacy of ketoconazole against leishmaniasis braziliensis panamensis cutaneous leishmaniasis. *Am J Med*, **89**, 147–155.
12. Navin TR, Arana BA, Arana FE, Berman JD, Chajéon (1992). Placebo-controlled clinical trial of sodium stibogluconate (Pentostam) versus ketoconazole for treating cutaneous leishmaniasis in Guatemala. *J Infect Dis*, **165**, 528–534.
13. Saenz RE, De Rodriguez CG (1991). Efficacy and toxicity of Pentostam against Panamanian mucosal leishmaniasis. *Am J Trop Med Hyg*, **44**, 394–398.
14. Franke ED, Wignall FS, Cruz ME, Rosales E, Tovar AA, Lucas CM, Llanos-Cuentas A (1990). Efficacy and toxicity of sodium antimony gluconate for mucosal leishmaniasis. *Ann Intern Med*, **113**, 934–940.
15. Rossi F, Acampora R, Vacca C, Maione S, Matera MG, Servodio R, Marmo E (1987). Prenatal and postnatal antimony exposure in rats: Effects on vasomotor reactivity development of pups. *Teratogenesis Carcinogen Mutagen*, **7**, 491–496.
16. Massip P, Goutner CH, Dupic Y, Navarrot P (1986). Kala-azar chez la femme enceinte. *La Presse Médicale*, **15**, 933.

17. Berman JD, Melby PC, Neva FA (1989). Concentration of Pentostam in human milk. *Trans R Soc Trop Med Hyg*, **83**, 784–785.
18. Herwaldt BL, Berman J (1992). Recommendations for treating leishmaniasis with sodium antimony gluconate (Pentostam) and review of pertinent clinical studies. *Am J Trop Med Hyg*, **40**, 296–306.
19. Veiga JPR, Wolff ER, Samoaio RNR, Marsden PD (1983). Renal tubular dysfunction in patients with mucocutaneous leishmaniasis treated with pentavalent antimonials. *Lancet*, **ii**, 569.
20. Jolliffe DS (1985). Nephrotoxicity of pentavalent antimonials. *Lancet*, **i**, 584.
21. Donovan KL, White AD, Cooke DA, Fisher DJ (1990). Pancreatitis and palindromic arthropathy with effusions associated with sodium stibogluconate treatment in a renal transplant recipient. *J Infect*, **21**, 107–110.
22. Martinez S, Looker DL, Berens RL, Marr JJ (1988). The synergistic action of pyrazolopyrimidines and pentavalent antimony against *Leishmania donovani* and *L. braziliensis*. *Am J Trop Med Hyg*, **39**, 250–255.
23. Martinez S, Marr J (1992). Allopurinol in the treatment of American cutaneous leishmaniasis. *N Engl J Med*, **326**, 741–744.
24. *WHO Model Prescribing Information. Drugs used in parasitic diseases* (1990). (Geneva: World Health Organization).

Artemisinin and its derivatives

Chemical structure

Artemisinin

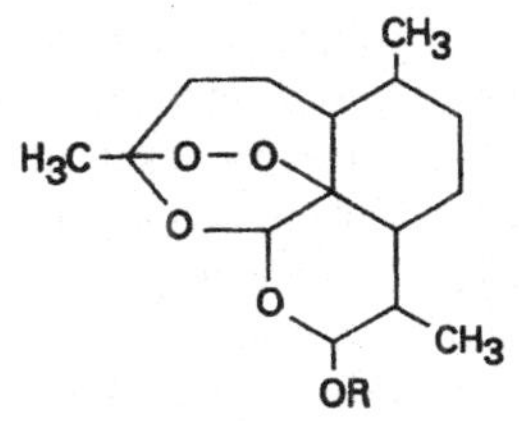

Artemether: R=CH_3

Arteether: R=CH_2CH_3

Artesunate: R=$C(=O)CH_2CH_2CO_2Na$

Physical properties

Artemisinin: MW 280; artesunate: MW 404; artemether: MW 296; arteether: MW 314. Artemisinin is poorly soluble in water, whereas its derivatives are more soluble. Artemether and artesunate are sensitive to moisture and acidic conditions. An aqueous solution of sodium artesunate of pH 7–8 hydrolyses rapidly to dihydroartemisinin.

Pharmacology and mechanism of action

Artemisinin (qinghaosu) is an antimalarial compound first isolated in pure form in 1972 by Chinese scientists from the herb qinghao (*Artemisia annua*). This herb (worm wood) has been used in Chinese traditional medicine to control fever for over 2000 years (1). Artemisinin is a compound with a peculiar structure, low toxicity and high efficacy even in severe chloroquine resistant *P. falciparum* malaria. Unlike current antimalarial drugs which have a nitrogen-containing heterocylic ring system, it is a sesquiterpene lactone with an endoperoxide linkage. The endoperoxide linkage is essential for the antimalarial activity of the drug. Artemisinin has been shown to be a potent schizontocidal drug both *in vitro* and in experimental animal models, but it has no practical effect against the exoerythrocytic tissue phase, the sporozoites and the gametocytes (2).

The mechanism of action of artemisinin is not clearly understood. The drug selectively concentrates in parasitized cells by reacting with the intraparasitic hemin (hemozoin). *In vitro* this reaction appears to generate toxic organic free radicals causing damage to parasite membranes (2–4). The derivatives of artemisinin are more potent than the parent drug and have apparently a similar mechanism of action (1, 2).

Pharmacokinetics

The assay of artemisinin and its derivatives in biological materials is extremely difficult. A number of HPLC methods have been published (5–9) but the sensitivity of these methods is generally unsatisfactory. For some of the methods the specificity can be questioned. Furthermore the artemisinin derivatives are strongly bound to erythrocytes (haemoglobin) and it has not been possible to determine the drug concentration in whole blood. The pharmacokinetic data for artemisinin and its derivatives are therefore limited.

Artemisinin can be given orally or rectally. Artesunate is given orally, intramuscularly or intravenously. Artemether is given orally or intramuscularly. Arteether is not yet available for use. Artemisinin and its derivatives seem to have similar pharmacokinetic profiles. After oral administration artemisinin is rapidly absorbed with peak plasma levels occurring within one hour (10). Relative bioavailability compared with an intramuscular oil injection was 32%. Rectal absorption of an aqueous suspension was poor and erratic compared with oral administration and intramuscular oil injection (10).

Artemisinin and its derivatives are strongly bound both to plasma proteins and to red blood cells (haemoglobin). Artemisinin, dihydroartemisinin, artemether and artesunate bind to different degrees to human serum proteins, particularly to alpha-acid glycoprotein; the rates of binding were found to be 64%, 43%, 76%, and 59%, respectively (1). Artemisinin and its derivatives are rapidly hydrolysed in the body to the active metabolite dihydroartemisinin which is mainly excreted via the urine in the form of metabolites (11, 12). Small amounts of the parent compounds may be excreted unchanged with the urine (11).

Recently, the pharmacokinetics of artemether was studied in healthy volunteers ($n = 6$) and in patients with uncomplicated malaria ($n = 8$) (13). After a single oral dose of 200 mg, average peak plasma levels of 118 ng/ml and 231 ng/ml respectively were reached about the same time after 3 hours. The metabolite (dihydroartemisinin) peak was also achieved after 3 hours. The mean ratio of metabolite to parent drug was 5:1 for the volunteers and 24:1 for the patients. Plasma elimination half-lives between 1–10 hours and 5–21 hours for the artemether and dihydroartemisinin respectively were estimated. These values reflect slow absorption rather than actual half-lives of the compounds. Large inter-individual variability in the plasma concentrations of the artemether and dihydroartemisinin was also observed which was likely to be due to differences in oral absorption.

Clinical trials

Artemisinin

Artemisinin and its derivatives have been used in China and Vietnam for a number of years. However, they are rapidly being introduced, officially or unofficially, in countries in Asia (Myanmar, Thailand), Africa (Tanzania, Malawi, Nigeria, Gambia and Sudan) and Latin America (Brazil) despite the fact that these compounds are still under clinical evaluation.

In the first documented report in English on the use of artemisinin, 1,511 patients with *P. vivax* and 588 patients with *P. falciparum* were clinically cured (defined in this instance as defervescence within 72 hours and clearance of parasitaemia within 120 hours after commencement of treatment) following a 3-day course of artemisinin given orally at a total dosage of 2.5–3.2 g intramuscularly in an oil solution, oil suspension or water suspension at total dosages of 0.5–0.8 g, 0.8–1.2 g and 1.2 g, respectively. No serious side effects have been observed during the treatment including patients with complicated heart, liver or renal diseases (1, 2). In comparative studies artemisinin cleared parasitaemia and fever more rapidly than chloroquine, quinine, mefloquine or a combination of mefloquine/sulphadoxine/pyrimethamine in Chinese (14–16) and Vietnamese (17, 18) patients with uncomplicated *falciparum* malaria. The total doses used in these studies varied from 0.6 g to 2.8 g for a duration of 2 to 3 days either orally, intramuscularly or rectally. In one study children were treated successfully with suppositories (16).

The most striking results from studies with artemisinin were the effects on chloroquine-resistant *falciparum* and complicated cerebral malaria. In 141 patients with cerebral malaria who were treated orally via a nasogastric tube or by intramuscular injection a mortality rate of only 7% was reported (2). In a similar study in children under 15 years a 9% mortality rate was reported (19). These figures are better than those reported for chloroquine or quinine in other studies. In a prospective randomized controlled study in patients with cerebral malaria in Vietnam, artemisinin suppositories were compared to artesunate and quinine (18). Artemisinin significantly increased initial parasite clearance, but did not reduce the mean coma duration time or mortality rate compared with quinine. However, artemisinin in suppository form was as effective as i.v. quinine.

One of the major problems with artemisinin or its derivatives is the high recrudescence rate (45–100%) which occurs within one month after treatment (20). Recrudescence may be linked to poor absorption of the drug in some individuals. In general the time effective inhibitory concentrations are present and might be insufficient for parasite eradication due to the short half-life and comparatively short treatment periods.

Artesunate

The data from 18 clinical studies on artesunate have recently been reviewed (12). In 4 of them (n = 109) artesunate was given parenterally for severe malaria. In 9 studies (n = 713) parenteral artesunate was given for uncomplicated malaria and in 5 (n = 272) artesunate was given orally in uncomplicated malaria. Eleven patients (10%) with severe malaria died but recovery was rapid in survivors; mean fever clearance times ranged between 30 and 40 hours and mean parasite clearance times between 28 and 55 hours. In uncomplicated malaria mean fever clearance times were between 14 and 38 hours and mean parasite clearance times between 17 and 68 hours. Recrudescence rates after a 3-day regimen were 49%. There was no local or systemic toxicity.

Artemether

The data of 19 clinical studies with artemether since 1982 have been reviewed recently (12). The studies included 812 patients with *falciparum* malaria with variable severity. Artemether was rapidly effective with mean fever clearance times of 17–47 hours (median 24 hours). In 14 studies fever clearance was more rapid in uncomplicated malaria (median: 22 hours; range: 17–30 hours) compared to 5 studies with severe malaria (median: 43 hours; range: 30–84 hours). There has been no evidence of significant systemic or local toxicity.

In two randomized studies intramuscular artemether was compared with intramuscular chloroquine or intravenous quinine in the treatment of complicated malaria in children in Africa. In Malawi (21) artemether (initial dose 3.2 mg/kg, then 1.6 mg/kg daily until recovery of consciousness) significantly reduced coma duration (8 *vs* 14 hours) and parasite clearance times (28 *vs* 48 hours) compared with quinine. The mortality rate was similar. In Gambia (22) artemether (initial dose 4 mg/kg then 2 mg/kg daily) was also associated with a significantly shorter time to parasite clearance than chloroquine (37 *vs* 48 hours) in 30 children with moderately severe malaria. Of the children treated with artemether 10% (2/22) died compared with 27% (6/22) mortality rate of the chloroquine group. No toxicity was recorded in either group.

Indications

Artemisinin and its derivatives are valuable drugs for the management of malaria. They should not be used unnecessarily or with incomplete dosage regimens. They are indicated only in areas where multidrug resistant *P. falciparum* malaria is prevalent (23).

Pregnancy and lactation

Artemisinin or its derivatives cause fetal resorption in rodents even at relatively low doses (above 10 mg/kg) when given after the sixth day of gestation (2). Experience in humans is still limited, particularly during early pregnancy. No ill effects have been reported in 23 children born to mothers given either artemisinin or artemether during the 16–38 week of pregnancy (23). Artemisinin or its derivatives should be given to pregnant women suffering from cerebral or complicated malaria in areas with multiresistant *P. falciparum*.

Excretion into breast milk is unknown.

Side effects

Artemisinin and its derivatives are exceptionally safe drugs. Millions of people have taken them and serious side effects have yet to be reported. The most commonly reported side effects include mild and transient gastrointestinal problems (such as nausea, vomiting, abdominal pain and diarrhoea), headache, and dizziness particularly after oral administration. Transient first degree heart block and bradycardia were reported in a few individuals, who received artesunate or artemether at the standard doses. Brief episodes of drug-induced fever have also been observed in a few studies (12, 23). After rectal administration the patients may experience tenesmus, abdominal pain and diarrhoea. A transient dose-related decrease in circulating reticulocytes has been reported following high doses of artesunate above 4 mg/kg for 3 days. All values returned to pre-treatment values within 14 days (12, 23). Neurotoxicity has been observed in animal studies but has never been documented in man (24).

Contraindications

There are no known contraindications. However, artemisinin and its derivatives should only be used when other antimalarial drugs do not work.

Drug interactions

There have been no reports.

Dosage (23)

In multidrug-resistant areas (adults and children over 6 months) the following apply.

Uncomplicated malaria

Artesunate (oral)
Day 1: 5 mg/kg as a single dose.
Day 2: 2.5 mg/kg as a single dose + Mefloquine 15–25 mg base/kg.
Day 3 2.5 mg/kg as a single dose.

Artemisinin (oral)
Day 1: 25 mg/kg as a single dose.
Day 2: 12.5 mg/kg as a single dose + Mefloquine 15–25 mg base/kg.
Day 3: 12.5 mg/kg as a single dose.

Severe and complicated malaria

Artemether (intramuscular)
3.2 mg/kg intramuscularly on the first day, followed by 1.6 mg/kg daily until the patient is able to take oral therapy of an effective antimalarial drug or to a maximum of 7 days. The drug can be given as a single daily injection. In children, the use of a 1 ml tuberculin syringe is advisable since the injection volumes will be small.

Artesunate (intramuscular or intravenous)
2 mg/kg on the first day, followed by 1 mg/kg/day until oral therapy is possible. In hyper-endemic areas, an alternative dose may be used: 2 mg/kg followed by 1 mg/kg 4–6 hours later then 1 mg/kg/day until oral therapy is possible.

Preparations

Artemether

- Paluther® (Rhône-Poulenc Rorer). Solution for injection 80 mg/ml.
- Artenam® (Dragon Pharmaceuticals Ltd, Wales UK). Solution for injection 100 mg/ml.

Several other preparations containing artemisinin derivatives are manufactured in China and Vietnam. The availability of these preparations is presently uncertain.

References

1. Luo XD, Shen CC (1987). The chemistry, pharmacology and clinical applications of qinghaosu (artemisinin) and its derivatives. *Med Res Rev*, **7**, 29–52.
2. Klayman DL (1985). Qinghaosu (artemisinin): an antimalarial drug from China. *Science*, **228**, 1049–1055.
3. Zhang F, Gosser Jr. DK, Meshnick SR (1992). Hemin-catalyzed decomposition of artemisinin (qinghaosu). *Biochem Pharmacol*, **43**, 1805–1809.
4. Meshnick SR, Yang YZ, Lima V, Kuypers F, Kamchonwongpaisan S, Yuthavong Y (1993). Iron-dependent free radical generation from the antimalarial artemisinin (qinghaosu). *Antimicrob Agents Chemother*, **37**, 1108–1114.
5. Zhao SS (1987). High performance liquid chromatographic determination of artemisinin (QHS) in human plasma and saliva. *Analyst*, **112**, 661–664.
6. Edlund PO, Westerlund D, Carlqvist J, Wu BL, Jin YH (1984). Determination of artesunate and dihydroartemisinin in plasma by liquid chromatography with post-column derivatization and UV-detection. *Acta Pharm Suec*, **21**, 223–234.
7. Thomas CG, Ward SA, Edwards G (1992). Selective determination, in plasma, of artemether and its major metabolite dihydroartemisinin by high-performance liquid chromatography with ultra-violet detection. *J Chromatogr*, **583**, 131–136.

8. Titulaer HAC, Vink-Blijleven N (1993). Assay of artelininc acid in serum by high-performance liquid chromatography. *J Chromatogr*, **612**, 331–335.
9. Idowu OR, Ward SA, Edwards G (1989). Determination of artelinic acid in blood plasma by high-performance liquid chromatography. *J Chromatogr*, **495**, 167–177.
10. Titulaer HAC, Zuidema J, Kager PF, Westeyn JCFM, Lugt CHB, Merkus FWHM (1990). The pharmacokinetics of artemisinin after oral intramuscular and rectal administration to human healthy volunteeers. *J Pharm Pharmacol*, **42**, 810–813.
11. Lee IS, Hufford CD (1993). Metabolism of antimalarial sesquiterpene lactones. *Pharmac Ther*, **48**, 345–355.
12. Hien TT, White NJ (1993). Qinghaosu. *Lancet*, **341**, 603–608.
13. Na Bangchang K, Karbwang J, Thomas CG, Thanavibul A, Sukontason K, Ward SA, Edwards G (1994). Pharmacokinetics of artemether after oral administration to healthy Thai males and patients with acute uncomplicated *falciparum* malaria. *Br J Clin Pharmacol*, **37**, 249–253.
14. Jiang JB, Li GQ, Guo XB, Kong YC, Arnold K (1982). Antimalarial activity of mefloquine and qinghaosu. *Lancet*, **2**, 285–288.
15. Li GQ, Arnold K, Guo XB, Jian HX, Fu LC (1984). Randomized comparative study of mefloquine qinghaosu and pyrimethamine-sulfadoxine in patients with *falciparum* malaria. *Lancet*, **2**, 1360–1361.
16. Hien TT, Tam DT, Cuc NT, Arnold K (1991). Comparative effectiveness of artemisinin suppositories and oral quinine in children with acute *falciparum* malaria. *Trans R Soc Trop Med Hyg*, **85**, 210–211.
17. Arnold K, Hien TT, Chinh NT, Phu NH, Mai PP (1990). A randomized comparative study of artemisinin (qinghaosu) suppositories and oral quinine in acute *falciparum* malaria. *Trans R Soc Trop Med Hyg*, **84**, 499–502.
18. Hien TT, Arnold K, Vinh H, Cuong BM, Phu NH, Chau TTH, Hoa NTM, Chuong LV, Mai NTH, Winh NN, Trang TTM (1992). Comparison of artemisinin suppositories with intravenous artesunate and intravenous quinine in the treatment of cerebral malaria. *Trans R Soc Trop Med Hyg*, **86**, 582–583.
19. Li GQ, Guo XB, Jin R, Wang ZC, Jian HX, Li ZY (1982). Clinical studies on the treatment of cerebral malaria with qinghaosu and its derivatives. *J Trad Chinese Med*, **2**, 125–130.
20. China Cooperative Research Group (1982) on Qinghaosu and its derivatives as antimalarials. Clinical studies on the treatment of malaria with qinghaosu and its derivatives. *J Trad Chinese Med*, **2**, 45–50.
21. Taylor TE, Wills BA, Kazembe P, Chisale M, Wirima JJ, Ratsma EY, Molyneux ME (1993). Rapid coma resolution with artemether in Malawian children with cerebral malaria. *Lancet*, **341**, 661–662.
22. White NJ, Waller D, Crawley J, Nosten F, Chapman D, Brewster D, Greenwood BM (1992). Comparison of artemether and chloroquine for severe malaria in Gambian children. *Lancet*, **339**, 317–321.
23. The role of artemisinin and its derivatives in the current treatment of malaria (1994–1995). Report of an informal consultation convened by WHO, 27–29 September, 1993. (Geneva: World Health Organization).
24. Brewer TG, Grate SJ, Peggins JO, Weina PJ, Petras JM, Levine BS, Heiffer MH, Schuster BG (1994). Fatal neurotoxicity of arteether and artemether. *Am J Trop Med Hyg*, **51**, 251–259.

Bephenium hydroxynaphthoate

Chemical structure

Bephenium hydroxynaphthoate

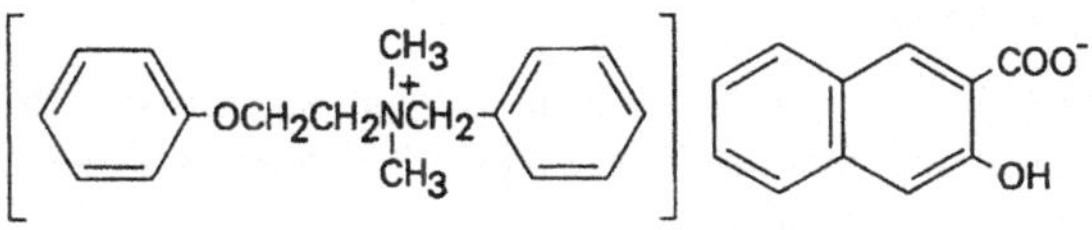

Physical properties

MW 256 (quaternary ammonium compound); bephenium hydroxynaphthoate: MW 444. It is practically insoluble in water. The drug should be kept in air-tight containers.

Pharmacology and mechanism of action

Bephenium is a quaternary ammonium compound first introduced into clinical medicine in 1958. It has a wide anthelminthic activity, in particular against *Ancylostoma duodenale* and *Ascaris lumbricoides*.

The mechanism of action of bephenium is similar to that of pyrantel and levamisole (see the monograph on levamisole, p. 74).

Pharmacokinetics

A specific analytical method has not been reported. The drug is poorly soluble and its absorption from the gastrointestinal tract is minimal. Less than 1% of the administered dose has been reported to be excreted with the urine in 24 hours (1).

Clinical trials

Early dose finding studies of the drug against ancylostomiasis and ascariasis have shown that a single dose above 2 g (base) was the optimal dose for both adults and children with egg reduction rates above 60% two weeks after treatment. Doses lower than this were ineffective and there was no substantial increase in the egg reduction rate after multiple dosage regimens. When the drug was given one hour after a saline purge, no increase of efficacy was reported. However, when the purge was given together with the drug, the results were less satisfactory (2). Studies conducted during the 1960s with a larger number of patients reported the drug to be highly effective against ancylostomiasis (cure rate between 80% and 100%) and ascariasis (cure rate between 50% and 80%) (3–6). In several open studies where the drug was compared to pyrantel and levamisole, it was equally effective against ancylostomiasis (cure rate close to 100%), but was less effective against ascariasis (cure rate around 80%) compared to the two drugs which have shown cure rates of around 100% for both parasites (7, 8, 9).

Bephenium hydroxynaphthoate has been reported to be less effective against *Necator americanus* (cure rate <50%) (10, 11).

Indications

Infections with *Ancylostoma duodenale*, *Ascaris lumbricoides* and *Necator americanus*. Superior alternative drugs are available today.

Pregnancy and lactation

Early clinical studies have reported the drug to be apparently safe during pregnancy (3, 12). Although modern documentation is lacking, the drug can be given during pregnancy if there is a strong indication for use.

Side effects

The drug has a bitter taste which might cause some patients especially children to refuse intake or vomit. This may be minimized by giving the drug with a sweet drink. Some individuals experience transient nausea, vomiting and headache after drug intake. Vertigo is occasionally reported.

Contraindications

There are no known contraindications to the drug.

Interactions

Bephenium has antagonistic effects to piperazine. However, no interactions were reported when the two drugs were combined together during helminthic therapy (2).

Dosage

Ancylostoma duodenale and *Ascaris lumbricoides*

Adults and children
5 g bephenium hydroxynaphthoate as a single dose.

Necator americanus

Adults and children >20 kg
5 g bephenium hydroxynaphthoate daily for 3 days.
Children <20 kg
Half the adult dose.

Bephenium is preferably given on an empty stomach.

Preparations

5 g bephenium hydroxynaphthoate contains 2.9 g bephenium.

- Alcopar® (Wellcome). Sachet containing 5 g of bephenium hydroxynaphthoate.

References

1. Rogers EW (1958). Excretion of bephenium salts in urine of human volunteers. *BMJ*, **II**, 1576–1577.
2. Goodwin LG, Jayewardene LG, Standen OD (1958). Clinical trials with bephenium hydroxy-naphthoate against hookworm in Ceylon. *BMJ*, **2**, 1572–1576.
3. Salem HH (1965). Clinical trials with bephenium hydroxynaphthoate against *Ancylostoma duodenale* and other intestinal helminths. *J Trop Med Hyg*, **68**, 21–25.
4. Abdalla A, Saif M (1963). The efficacy of single-dose treatment of ancylostomiasis with bephenium hydroxynaphthoate. *J Trop Med Hyg*, **66**, 45–47.
5. Hsieh H-C, Brown HW, Fite M, Chows L-P, Cheng C-S, Hsu C-C (1960). The treatment of hookworm, *Ascaris* and *Trichuris* infections with bephenium hydroxynaphthoate. *Amer J Trop Med Hyg*, **9**, 496–499.
6. Hahn SS, Kang HY, Hahn YS (1960). The anthelminthic effect of bephenium hydroxynaphthoate on intestinal helminths. *J Trop Med Hyg*, **63**, 180–184.
7. Farahmandian I, Arfaa F, Jalali H, Reza M (1977). Comparative studies on the evaluation of the effect of new anthelminthics on various intestinal helminthiasis in Iran. *Chemotherapy*, **23**, 98–105.
8. Farid Z, Bassily S, Miner WF, Hassan A , Laughlin LW (1977). Comparative single doses treatment of hookworm and roundworm infections with levamisole, pyrantel and bephenium. *J Trop Med Hyg*, **80**, 107–108.
9. Al-Issa TB, Abdul Wahab H (1977). Comparative trial of pyrantel, levamisole and bephenium in the treatment of intestinal worms in Iraq. *Bull Endem Dis*, **18**, 109–115.
10. Nahmias J, Kennet R, Goldsmith R, Greenberg Z (1989). Evaluation of albendazole, pyrantel, bephenium, pyrantel-praziquantel and pyrantel-bephenium for single-dose mass treatment of necatoriasis. *Ann Trop Med Parasitol*, **83**, 625–629.
11. Hettiarachchi J, Senewiratne K (1975). A comparative study of the relative efficacy of pyrantel pamoate, bephenium hydroxynaphthoate and tetrachlorethylene in the treatment of *Necator americanus* infection in Ceylon. *Ann Trop Med Parasitol*, **69**, 233–239.
12. Davidson JC (1962). The treatment of hookworm infection with bephenium hydroxynaphthoate. *Centr Afr Med J*, **8**, 272.

Bithionol

Chemical structure

Bithionol

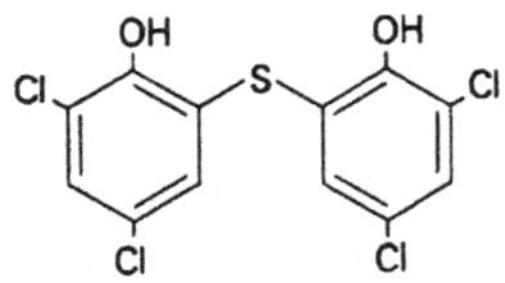

Physical properties

MW 356; pKa not known. Practically insoluble in water.

Pharmacology and mechanism of action

Bithionol is a dichlorophenol which is structurally related to hexachlorophene and niclosamide. It was introduced into clinical medicine three decades ago. The drug has been replaced by praziquantel, but it is still used in some areas of the world, particularly against *Fasciola hepatica*.

The mechanism of action of bithionol is not well known. It inhibits oxidative phosphorylation of *Paragonimus westermani* and causes morphological alterations in *Fasciola hepatica* (1, 2, 3).

Pharmacokinetics

Specific analytical methods have not been described.

The drug is taken only orally. It is readily absorbed from the gastrointestinal tract, glucuronidated in the liver and excreted via the kidneys (4).

Clinical trials

In an open study, bithionol 40 mg/kg on alternate days for a total of 15 doses was given to 1355 patients with pulmonary paragonimiasis. After one year follow-up, a cure rate of 90% was reported (5). In a similar study, where 40 mg/kg was given daily to 39 patients with pulmonary paragonimiasis for 15 to 25 doses, all but one, who relapsed 6 months after treatment, were cured. The relapsing patient was apparently cured by a second course of bithionol (6). Less satisfactory results have been reported by others (7).

Bithionol at a dosage of 25 mg/kg daily for 10 days was given to 10 patients with *Fasciola hepatica* (8 with acute fascioliasis). All patients were cured. However, 3 patients required a further dose after 2 to 3 months due to relapse. The follow-up period lasted 16 to 47 months (8). In two other studies with patients with *Fasciola hepatica*, a cure rate of 100% was reported after giving 5 doses of bithionol 30 mg/kg each every other day. No relapse has been reported after a follow-up period of 3 to 4 months. The number of patients included in the two studies were 8 and 14, respectively (9, 10). In another open study with a limited number of patients with fascioliasis, the efficacy of praziquantel (75 mg/kg given in 3 daily divided doses for 10 days), dehydroemetine (1 mg/kg i.m. daily for 10 to 14 days) and bithionol (40 mg/kg on alternate days for 14 days) was compared. Praziquantel and bithionol were less effective than dehydroemetine. Of 8 patients treated with bithionol, only 3 patients were reported to have been cured. The other 5 patients were treated with dehydroemetine. The length of the follow-up period was not reported (11).

Indications

Infections with *Parogonimus westermani* (lung fluke) and *Fasciola hepatica* (liver fluke).

Bithionol should only be used as a second-line drug in the treatment of paragonimiasis for those patients who fail to respond to full course therapy with praziquantel. Triclabendazole, which is a new drug still under clinical evaluation, will most probably become the drug of choice against fascioliasis in the near future.

Pregnancy and lactation

Documentation is lacking both in animals and in man. Bithionol should not be used during early pregnancy unless there is a strong indication for use.

Its excretion into breast milk is not known.

Side effects

Side effects are common, but minor. About one third of the patients may experience diarrhoea that may be accompanied by anorexia, nausea and vomiting. Skin rashes or urticaria usually together with itching may be seen. Phototoxic reactions can occur (1, 5, 12, 13).

Contraindications and precautions

Efficacy and safety of the drug has not been established in children under 8 years of age.

Interactions

There have been no reports.

Dosage

Paragonimus westermani and *Fasciola hepatica*

Children and adults
30–50 mg/kg orally every other day in two divided doses to a total of 10–15 doses.

Preparations

- Bitin® (Tanabe Seiyaky) Tablets 500 mg.

References

1. Barret-Connor E (1982). Drugs for the treatment of parasitic infection. *Med Clin North Am*, **66**, 245–255.
2. Dawes B. Some apparent effects of Bithionol (Actamer) on *Fasciola hepatica*. *Nature*. **209**, 424–425.
3. Yokogawa M, Muneo I (1965). Paragonimiasis. *Adv Parasitol*, **3**, 99–158.
4 Takada M (1976). On the metabolic detoxication of bithionol in man. *J Toxicol Sci*, **1**, 26–31.
5. Chung HL, Ho LY, Hsu CP, Ts'ao WJ (1981). Recent progress in studies of *Paragonimus* and paragonimiasis control in China. *Chin Med J*, **94**, 493–494.
6. Singh TS, Mutum SS, Razaque MA (1986). Pulmonary paragonimiasis: Clinical features, diagnosis and treatment of 39 cases in Manipur. *Trans R Soc Trop Med Hyg*, **80**, 967–971.
7. Oh SJ (1967). Bithionol treatment in cerebral paragonimiasis. *Am J Trop Med Hyg*, **16**, 585–590.
8. Bacq Y, Besnier JM, Duong TH, Pavie G, Metman ET, Choutet P (1991). Successful treatment of acute fascioliasis with bithionol. *Hepatology*, **14**, 1066–1069.
9. Farag HF, Salem A, El-Hifni SA, Kandil M (1988). Bithionol (Bitin) treatment in established fascioliasis in Egyptians. *J Trop Med Hyg*, **91**, 240–244.
10. Bassiouny HK, Soliman NK, El-Daly SM, Badr NM (1991). Human fascioliasis in Egypt: effect of infection and efficacy of bithionol treatment. *J Trop Med Hyg*, **94**, 333–337.
11. Farid Z, Kamal M, Woody J (1988). Treatment of acute toxemic fascioliasis. *Trans R Soc Trop Med Hyg*, **82**, 299.
12. Kim JS (1970). Treatment of *Paragonimus westermani* infections with Bithionol. *Am J Trop Med Hyg*, **19**, 940–942.
13. O'Quinn SE, Kennedy CB, Isbell KH (1967). Contact photodermatitis due to bithionol and related compounds. *J Am Med Ass*, **199**, 89–92.

Chloroquine

Chemical structure

Chloroquine

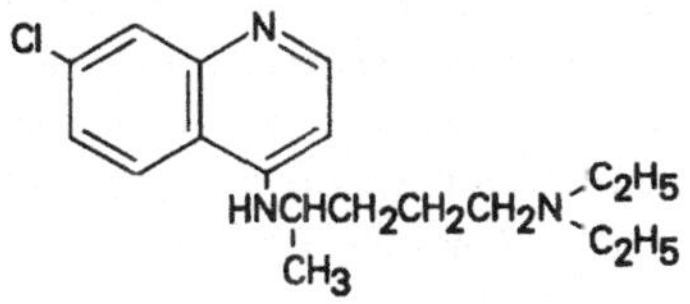

Physical properties

Base: MW 320; phosphate: MW 516. 1 g dissolves in 4 ml of water. Sulphate: MW 436; pKa 8.4, 10.8. 1 g dissolves in 3 ml of water. The drug should be protected from light.

Pharmacology and mechanism of action

Chloroquine was first synthesized by Andersag in 1934 and the initial clinical studies were performed in Germany. The drug later underwent extensive clinical studies by the Americans in 1944 and was found to be an outstanding antimalarial compound. Chloroquine is a racemate with two enantiomers with similar antimalarial activity. It is a potent schizontocidal drug which is highly effective against the asexual forms of all four species of malaria. It is also active against gametocytes of *Plasmodium (P.) vivax*, *P. malariae* and *P. ovale*, but not against *P. falciparum* (1).

Development of chloroquine-resistant *P. falciparum* parasites is now widespread. Much work has been done to understand the mechanisms of this resistance. It is now well documented that red blood cells infected with resistant plasmodia concentrate less chloroquine than those infected with susceptible strains (2, 3). This is due to an increased efflux of chloroquine from chloroquine-resistant parasites (4). It has also been demonstrated that the calcium channel blocker verapamil and a number of other drugs can inhibit this efflux and reverse chloroquine resistance *in vitro* (5, 6), but the doses of these inhibitors used *in vitro* are too high to be clinically useful. Its metabolite deethylchloroquine is less active *in vitro* against resistant *P. falciparum* strains, but not against susceptible strains (7).

The mechanism of action of chloroquine is probably related to its inhibition of the enzyme that polymerizes and detoxifies ferriprotoporphyrin IX in the parasite food vacuole (8).

Chloroquine has also amoebicidal and anti-inflammatory properties (9).

Pharmacokinetics

Earlier pharmacokinetic studies are based on unspecific fluorescence methods (10, 11). HPLC, GC and radioimmunoassay analytical methods have been described for the determination of chloroquine and its metabolites. HPLC methods are commonly used (12, 13, 14).

Chloroquine binds to blood constituents such as thrombocytes and granulocytes to such an extent that the plasma concentrations of the drug are only about 15% of that in whole blood (15). It is released from platelets during clotting and determinations of the drug in serum should generally be avoided (15, 16). Simple and accurate methods (HPLC) for capillary blood samples dried on filter paper strips are described (17, 18).

Chloroquine is given orally, intramuscularly, intravenously, rectally or by nasogastric administration. The oral bioavailability is approximately 90% (19), but has been reported to be significantly reduced in children with kwashiorkor (20). Chloroquine is also well absorbed when given intramuscularly or after nasogastric administration (21).

When given rectally to adult volunteers the bioavailability was only 22–24% of that after oral dosing (22, 23). However, rectal administration of non-coated chloroquine tablets to children with malaria gave concentrations that were marginally lower than after oral intake and in the absence of alternatives rectal administration might be tried (24).

Following a single oral dose of 300 mg base peak plasma levels of 60–90 ng/ml and 10–20 ng/ml for chloroquine and the metabolite deethylchloroquine respectively are reached within 1–6 hours. Chloroquine has a large volume of distribution of about 200 l/kg (19) and therefore haemo- and peritoneal dialysis has no place in the management of chloroquine intoxication. Chloroquine binds with high affinity to melanin-containing tissues such as the retina, the inner ear and hair follicles (25). In plasma about 46–74% of the drug is bound to plasma proteins mainly to albumin and α-acid glycoprotein (26).

Chloroquine is eliminated slowly from the body and it can be detected in the urine for more than a year after intake (27). It has a multiexponential elimination pattern with an initial elimination phase with a half-life of 3–6 days followed by a slower phase with a half-life of 12–14 days. Terminal elimination half-lives of chloroquine and deethylchloroquine of up to 2 months have been reported (27). However, this half-life is of minor importance for the overall disposition of the drug and its metabolite in the body. At steady state during malaria prophylaxis a mean half-life of 4.5 days was found for chloroquine (28). A dose-dependent elimination of chloroquine has been reported (29, 30), but the studies were made with unsatisfactory methodology and others were unable to confirm the results (31, 32).

Different studies using either single or repeated daily doses reported urinary recoveries of 45–56% of the dose within 3–13 weeks. About 70% of this was accounted for by the parent drug and 23% by deethylchloroquine. A further 8–10% is probably eliminated with the faeces (19, 33).

Clinical efficacy

Chloroquine resistance was first reported from Colombia in 1961 (34). Soon afterwards a second focus of resistance was reported in Thailand (35). The resistance in Southeast Asia increased rapidly and in the beginning of the 1970s chloroquine was ineffective and was abandoned for the treatment of *P. falciparum* malaria in Thailand (36).

P. falciparum remained highly susceptible to chloroquine in tropical Africa for a long time, but in 1978 resistance was reported from East Africa (37). From this initial focus in Kenya–Tanzania, chloroquine resistance has spread throughout the continent and has now been reported from all countries south of the Sahara (38).

Before chloroquine resistance became apparent in Africa, a single dose of 10 mg base/kg was effective in eradicating *P. falciparum* parasites in semi-immune individuals. Now 10 mg base/kg is no longer effective in the indigenous population, and a total dose of

25 mg base/kg is necessary for successful therapy (39). With a further increase in resistance, an increase in dose to 35–50 mg base/kg over 4–5 days (40) in Burundi and 30 mg base/kg (including a divided loading dose of 20 mg base/kg during the first 12 hours) in Madagascar had a better effect than the conventional dose of 25 mg base/kg. However, chloroquine has a narrow therapeutic range and doses exceeding a total of 50 mg base/kg over a few days are likely to result in concentration dependent adverse reactions in some individuals (19).

In 1990 in areas in Kenya and Malawi, standard chloroquine treatment (total 25 mg base/kg) to children under five years of age failed to produce either a durable clinical improvement or optimal haematological recovery (41). The response to sulphadoxine/pyrimethamine was significantly better and Malawi has become the first sub-Saharan country to abandon chloroquine as the first line treatment of *falciparum* malaria in young children.

The efficacy of chloroquine prophylaxis depends on the degree of resistance. Unfortunately there are no large prospective studies comparing the efficacy of different prophylactic regimens. In areas with either predominantly *P. vivax* malaria or very low risk for *P. falciparum* malaria, chloroquine can still be used as monoprophylaxis. In tropical Africa the malaria transmission is much higher than in other parts of the world and *P. falciparum* malaria predominates. Therefore chloroquine has to be combined with proguanil for visitors to this area.

P. vivax with decreased susceptibility to standard chloroquine treatment has been reported from Papua New Guinea in 1989 (42). A prospective study done in 1991 in the Indonesian part of New Guinea demonstrated both a high failure rate in small children and breakthroughs during standard prophylaxis (43). There are no reports about decreased chloroquine sensitivity for *P. ovale* or *P. malariae*.

Indications

Despite the presence of resistant *P. falciparum* parasites, chloroquine is still with few exceptions, the drug of choice for treatment of non-severe malaria in the semi-immune indigenous population in Africa south of the Sahara. It should, however, not be used in severe or complicated malaria. In areas with considerable risk for chloroquine resistant *P. falciparum*, the drug has to be combined with another antimalarial drug, e.g. proguanil, for effective prophylaxis. Chloroquine still remains the drug of choice for treatment and prophylaxis of malaria caused by the other forms of plasmodia except in New Guinea.

Chloroquine is also useful for the treatment of amoebic hepatitis and in rheumatoid arthritis.

Pregnancy and lactation

Chloroquine is not teratogenic in man with the recommended prophylactic and treatment doses used against malaria (44). It should be used freely during pregnancy in areas with chloroquine sensitive *P. falciparum*.

Chloroquine is excreted into breast milk, but the amount is not sufficient to protect the infant from malaria (9).

Side effects

The dosage used during standard treatment and prophylaxis of malaria is usually well tolerated. Minor side effects such as nausea, vomiting, diarrhoea, dizziness, headache, rash and abdominal pain may occur. Accommodation problems at the time of maximal concentration a few hours after intake can be disturbing but disappear rapidly. Many of the mild concentration dependent side effects can be avoided if chloroquine is taken with some food at bedtime (9). Itching is common in blacks (45, 46, 47) and has been reported in Caucasians as well (48). It is a subjective reaction characterized by a widespread pricking sensation affecting mainly the hands, feet and the scalp without any rash. It begins 6–12 hours after drug intake and usually subsides within 3–4 days after stopping treatment. Chloroquine has been reported to exacerbate psoriasis (49). Acute cardiovascular effects have been reported to be associated with parenteral administration of chloroquine, (50) but can be avoided by slow intravenous infusion (51).

Chloroquine is known to carry a risk for toxic retinopathy when used in high doses for treatment of rheumatoid arthritis, particularly if the cumulative dose is above 100 g (52). In a recent study of 588 missionaries who have taken a median cumulative dose of chloroquine of more than 300 g for malaria prophylaxis (some of them for up to 24 years), none of the subjects were reported to have retinopathy (53).

Rare side effects occasionally reported include photosensitivity, tinnitus, reduced hearing and deafness, neuromyopathy, involuntary movements and aplastic anaemia, agranulocytosis, thrombocytopenia and neutropenia (52).

Chloroquine is a well established but rare cause of toxic psychosis (54–56). The symptoms disappear within two weeks of drug discontinuation. An association between epileptic seizures and chloroquine use has been reported (57). During long-term treatment (e.g. of rheumatoid arthritis) with high doses of chloroquine side effects are more common. Most of them are related to the central nervous system and include diplopia, blurred vision, accommodation problems, apathy, anxiety, fatigue, headache, dizziness and vertigo. These side effects are probably concentration dependent (19, 29).

Contraindications and precautions

Chloroquine should not be given to persons with a history of epilepsy or psychosis. It should also be avoided in psoriatic patients. Since chloroquine is mainly excreted unchanged by the kidney, dosage reductions may be considered in patients with kidney impairment. Cardiovascular collapse may occur after parenteral administration of chloroquine in severely ill patients especially in children. The drug, must therefore be given as a slow intravenous infusion with adequate fluid replacement (see below).

Interactions

Concurrent chloroquine intake may increase the mouth ulcerations reported with proguanil (58). Cimetidine inhibits the metabolism of chloroquine and may cause increased plasma levels of the latter (59).

Dosage

The dose of chloroquine is always calculated in terms of the base.

Treatment of malaria

A. Oral administration

Adults and children

A total dose of 25 mg base /kg is given over 3 days. The drug may also be given by nasogastric intubation.

Day 1: 15 mg base/kg (10 mg/kg as first dose followed by 5 mg/kg 6 hours later).
Day 2–3: 5 mg base/kg daily as a single dose.

B. Parenteral administration
Normally, chloroquine is given orally, but in patients who are unable to take drugs orally chloroquine can be given parenterally (51). Due to the existence of chloroquine resistant *P. falciparum* parasites in all endemic areas, chloroquine should generally not be used for treatment of severe or complicated *falciparum* malaria, particularly in non-immune subjects.

Adults and children
An initial dose of 10 mg base/kg should be administered over a period of 8 hours preferably by slow intravenous infusion. Rapid infusion or i.v. injection must be avoided due to the risk for cardiotoxicity (51). Subsequent infusions of 5 mg/kg should be administered every 8 hours until a total dose of 25 mg base/kg has been given. The administration should be switched to oral as soon as possible. Where facilities for intravenous infusion are not available, chloroquine can be administered by intramuscular or subcutaneous injection at a dosage of 3.5 mg base/kg every 6 hours until a total of 25 mg base/kg has been given.

Malaria prophylaxis

Adults including pregnant women
5 mg base/kg weekly or:

<70 kg	300 mg base
71–89 kg	375 mg base
90–105 kg	450 mg base

Children
5 mg base/kg weekly.

Treatment of amoebic hepatitis

Adults
600 mg base (or 10 mg/kg) daily for 2 days followed by 300 mg base (or 5 mg/kg) for another 2–3 weeks.

Children
10 mg base/kg daily (max. 300 mg base) for 2–3 weeks.

Preparations

Numerous preparations (tablets, oral solutions, solutions for injection) containing chloroquine phosphate or sulphate are available. 161 mg chloroquine phosphate equals 100 mg chloroquine base. 136 mg chloroquine sulphate equals 100 mg chloroquine base.

References

1. Black RH, Canfield CJ, Clyde DF, Peters W, Wernsdorfer WH (1986). Quinine. In: *Chemotherapy of Malaria* 2nd edn, edited by L Bruce-Chwatt. Monograph series no. 27. (Geneva: World Health Organization).
2 Fitch CD (1969). Chloroquine resistance in malaria. A deficiency of chloroquine binding. *Proceedings of the National Academy of Science of the USA*, **64**, 1181–1187.

3. Verdier F, Le Bras J, Clavier F, Hatin I, Blayo MC (1985). Chloroquine uptake by *Plasmodium falciparum*-infected human erythrocytes during *in vitro* culture and its relationship to chloroquine resistance. *Antimicrob Agents Chemother*, **27**, 561–564.
4. Krogstad DJ, Gluzman IY, Kyle DE, Odoula AMJ, Martin SK, Milhous WK, Schlesinger PH (1987). Efflux of chloroquine from *Plasmodium falciparum*: mechanism of chloroquine resistance. *Science*, **238**, 1283–1285.
5. Martin SK, Oduola AMJ, Milhous WK (1987). Reversal of chloroquine resistance in *Plasmodium falciparum* by verapamil. *Science*, **235**, 899–901.
6. Bitonti AJ, Sjoerdsma A, McCann PP, Kyle DE, Oduola AMJ, Rossan RN, Milhous WK, Davidsson DE (1988). Reversal of chloroquine resistance in malaria parasite *Plasmodium falciparum* by desipramine. *Science*, **242**, 1301–1303.
7. Aderounmu AF, Fleckenstein L (1983). Pharmacokinetics of chloroquine diphosphate in the dog. *J Pharmacol Exp Ther*, **226**, 633–639.
8. Slater AFG, Cerami A (1992). Inhibition by chloroquine of a novel haem polymerase enzyme activity in malaria trophozoites. *Nature*, **355**, 167–169
9. *Martindale: The Extra Pharmacopoeia*, 30th edn, (1993). Chloroquine and other antimalarials (London: Pharmaceutical Press), pp. 394–397.
10. Brodie BB, Udenfriend S, Dill W, Chenkin T (1947). The estimation of basic organic compounds in biological material. III Estimation by conversion to fluorescent compounds. *J Biol Chem*, **168**, 315–325.
11. McChesney EW, Wyzan HS, McAuliff JP (1956). The determination of 4-aminoquinoline antimalarials: Reevaluation of the induced fluorescence method with specific application to hydroxychloroquine analysis. *J Am Pharm Ass*, **45**, 640–645.
12. Alván G, Ekman L, Lindström B (1982). Determination of chloroquine and its desethyl metabolite in plasma red blood cells and urine by liquid chromatography. *J Chromatogr*, **229**, 241–247.
13. Bergqvist Y, Frisk-Holmberg M (1980). Sensitive method for the determination of chloroquine and its metabolite desethylchloroquine in human plasma and urine by high performance liquid chromatography. *J Chromatogr*, **221**, 119–127.
14. Brown ND, Poon BT, Chulay JD (1982). Determination of chloroquine and its de-ethylated metabolites in human plasma by ion-pair high-performance liquid chromatography. *J Chromatogr*, **229**, 248–254.
15. Bergqvist Y, Domeij-Nyberg B (1983). Distribution of chloroquine and its metabolite desethylchloroquine in human blood cells and its implication for the quantitative determination of these compounds in serum and plasma. *J Chromatogr*, **272**, 137–148.
16. Rombo L, Ericsson Ö, Alván G, Lindström B, Gustafsson LL, Sjöqvist F (1985). Chloroquine and desethylchloroquine in plasma serum and whole-blood – problems in assay and handling of samples. *Ther Drug Monit*, **7**, 211–215.
17. Lindström B, Ericsson Ö, Alván G, Rombo L, Ekman L, Rais M, Sjöqvist F (1985). Determination of chloroquine and its desethyl metabolite in whole blood. An application for samples collected in capillary tubes and dried on filter paper. *Ther Drug Monitor*, **7**, 207–210.
18. Patchen LC, Mount DL, Schwartz IK, Churchill FC (1983). Analysis of filter-paper-absorbed finger-stick blood samples for chloroquine and its major metabolite using high performance liquid chromatography with fluorescence detection. *J Chromatogr*, **278**, 81–89.
19. Gustafsson LL, Walker O, Alván G, Beermann B, Estevez F, Gleisner L, Lindström B, Sjöqvist F (1983). Disposition of chloroquine in man after single intravenous and oral doses. *Br J Clin Pharmacol*, **15**, 471–479.
20. Walker O, Dawodu AH, Salako LA, Alván G, Johnson AOK (1987). Single dose disposition of chloroquine in kwashiorkor and normal children – evidence for decreased absorption in kwashiorkor. *Br J Clin Pharmacol*, **23**, 467–472.
21. White NJ, Miller K, Churchill FC, Berry C, Brown J, Williams SB, Greenwood BM (1988). Chloroquine treatment of severe malaria in children. Pharmacokinetics, toxicity and new dosage recommendations. *N Engl J Med*, **319**, 1494–1500.
22. Minker E, Iván J (1991). Experimental and clinicopharmacological study of rectal absorption of chloroquine. *Acta Physiol Hung*, **77**, 237–248.
23. Tjoeng MM, Hogeman PHG, Kapelle H, De Ridder MLJ, Verhaar H (1991). Comparative bioavailability of rectal and oral formulations of chloroquine. *Pharm Weekbl Sci*, **13**, 176–78.
24. Westman L, Kamanda S, Hellgren U, Ericsson Ö, Rombo L (1994). Rectal administration of chloroquine for treatment of children with malaria. *Trans R Soc Trop Med Hyg*, **88**, 446.

25. Lindquist NG (1973). Accumulation of drugs on melanin. *Acta Radiol*, **325**, 1–9.
26. Walker O, Birkett DJ, Alván G, Gustafsson LL, Sjöqvist F (1983). Characterization of chloroquine plasma protein binding in man. *Br J Clin Pharmacol*, **15**, 375–377.
27. Gustafsson LL, Lindström B, Grahnén A, Alván G (1987). Chloroquine excretion following malaria prophylaxis. *Br J Clin Pharmacol*, **29**, 221–224.
28. Rombo L, Bergqvist Y, Hellgren U (1987). Chloroquine and desethyl-chloroquine concentrations during regular long-term malaria prophylaxis. *Bull World Health Organ*, **65**, 879–83.
29. Frisk-Holmberg M, Bergqvist Y, Domeij-Nyberg B, Hellström L, Jansson F (1979). Chloroquine serum concentration and side effects: evidence for dose-dependent kinetics. *Clin Pharmacol Ther*, **25**, 345–350.
30. Frisk-Holmberg M, Bergqvist Y, Termond E (1985). Further support for changes in chloroquine disposition and metabolism between a low and a high dose. *Eur J Clin Pharmacol*, **28**, 721–722.
31. Gustafsson LL, Rombo L, Alván G, Björkman A, Lind M, Walker O (1983). On the question of dose-dependent chloroquine elimination of a single oral dose. *Clin Pharmacol Ther*, **34**, 383–385.
32. Tett SE, Cutler DJ (1987). Apparent dose-dependence of chloroquine pharmacokinetics due to limited assay sensitivity and short sampling times. *Eur J Clin Pharmacokinet*, **31**, 729–731.
33. McChesney EW, Fasco MJ, Banks WF Jr (1967). The metabolism of chloroquine in man during and after repeated oral dosage. *J Pharmacol Exp Ther*, **158**, 323–331.
34. Moore DV, Lanier JE (1961). Observation on two *Plasmodium falciparum* infections with an abnormal response to chloroquine. *Am J Trop Med Hyg*, **10**, 5–9.
35. Harinasuta T, Migasen S, Bunnag D (1962). Chloroquine resistance in *Plasmodium falciparum* in Thailand. UNESCO 1st regional symposium on scientific knowledge of tropical parasites 5–9 November, 1962. University of Singapore, Singapore, pp. 148–153.
36. Chongsuphajaisiddhi T, Sabachareon A, Puangpartk S, Harinasuta T (1979). Treatment of *falciparum* malaria in Thai children. *Southeast Asian J Trop Med Public Health*, **10**, 132–137.
37. Centers for Disease Control (1978). Chloroquine-resistant malaria acquired in Kenya and Tanzania – Denmark, Georgia, New York. *Morbidity and Mortality Weekly Report*, **27**, 463–464.
38. *International Travel and Health* (1994) (Geneva: World Health Organization).
39. Onori E, Payne D, Grab B, Horst HI, Franco A, Joia H (1982). Incipient resistance of *Plasmodium falciparum* to chloroquine among a semi-immune population of the United Republic of Tanzania. I. results of *in vivo* and *in vitro* studies and of an ophtalmological survey. *Bull World Health Organ*, **60**, 77–87.
40. Coosemans MH, Hendrix L, Barutwanayo M, Butoyi G, Onori E (1985). Drug resistance of *Plasmodium falciparum* in Burundi. *Bull World Health Organ*, **63**, 331–338.
41. Bloland PB, Redd SC, Kazembe P *et al.* (1991). Co-trimoxazole for childhood febrile illness in malaria-endemic regions. *Lancet*, **337**, 518–520.
42. Rieckmann KH, Davis DR, Hutton DC (1989). *Plasmodium vivax* resistance to chloroquine? *Lancet*, **ii**, 1183–1184.
43. Murphy G, Basri H, Purnomo, Andersen EM, Bangs MJ, Mount DL, Gorden J *et al.* (1993). *Vivax* malaria resistant to treatment and prophylaxis with chloroquine. *Lancet*, **341**, 96–100.
44. Wolfe MS, Cordero JF (1985). Safety of chloroquine in chemosuppression of malaria during pregnancy. *BMJ*, **290**, 1466–1467.
45. Olatunde A (1977). The practical and therapeutic implications of chloroquine-induced itching in tropical Africa. *Afr J Med Sci*, **6**, 27–31.
46. Ajayi AA, Oluokun A, Sofowora O, Akinleye A, Ajayi AT (1989). Epidemiology of antimalarial-induced pruritus in Africans. *Eur J Clin Pharmacol*, **37**, 539–540.
47. Sowunmi A, Walker O, Salako LA (1989). Pruritus and antimalarial drugs in Africans. *Lancet*, **2**, 213.
48. Poulter NR, Lury JD, Poulter CJ (1982). Chloroquine-associated pruritus in a European. *BMJ*, **285**, 1703–1704.
49. Schopf RE, Ockenfels HM, Schultewolter T, Morsches B (1983). Chloroquine stimulates the mitogen-driven lymphocyte proliferation in patients with psoriasis. *Dermatology*, **187**, 100–103.
50. Looareesuwan S, White NJ, Chanthavanich P, Edwards G, Nicholl DD, Bunch C, Warrell DA (1986). Cardiovascular toxicity and distribution kinetics of intravenous chloroquine. *Br J Clin Pharmacol*, **22**, 31–36.
51. White NJ, Watt G, Bergqvist Y, Njelesani EK (1987). Parenteral chloroquine for treating *falciparum* malaria. *J Infect Dis*, **155**, 192–201.

52. Tester-Dalderup CBM (1992). Antiprotozoal drugs: Antimalarial drugs. Meyler's *Side Effects of Drugs*, 12th edn, pp 688–692.
53. Lange WR, Frankenfield DL, Moriarty-Sheehan M, Contoreggi CS, Frame JD *et al.* (1994). No evidence for chloroquine-associated retinopathy among missionaries on long-term malaria prophylaxis. *Am J Trop Med Hyg*, **51**, 389–392.
54. Brokes DB (1966). Chloroquine psychosis. *BMJ*, **I**, 983.
55. Dornhorst AC, Robinson BF (1963). Chloroquine psychosis? *Lancet*, **i**, 108.
56. Good MI, Shader RI (1977). Behavioral toxicity and equivocal suicide associated with chloroquine and its derivatives. *Am J Psychiatry*, **134**, 798–801.
57. Fish DR, Espir MLE (1988). Convulsions associated with prophylactic antimalarial drugs: implications for people with epilepsy. *BMJ*, **297**, 526–527.
58. Drysdale SF, Phillips-Howard PA, Behrens RH (1990). Proguanil, chloroquine, and mouth ulcers. *Lancet*, **335**, 164.
59. Ette EI, Brown-Awala EA, Essien EE (1987). Chloroquine elimination in humans: effect of low-dose cimetidine. *J Clin Pharmacol*, **27**, 813–881.

Dehydroemetine

Chemical structure

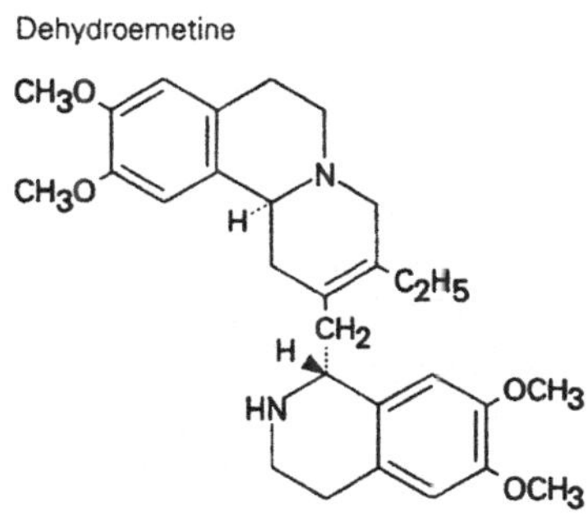

Physical properties

Base: MW 479; hydrochloride: MW 552; pKa not known. 1 g dissolves in 30 ml of water.

Pharmacology and mechanism of action

Dehydroemetine, is a derivative of emetine and has similar pharmacological properties, but it is considered to be less toxic. Dehydroemetine acts against *Entamoeba histolytica* in tissue and has little effect on amoeba confined to the lumen of the bowel. It has a direct amoebicidal effect *in vitro* by interfering with the mRNA translocation along the ribosome, which causes inhibition of protein synthesis (1).

Pharmacokinetics

A specific analytical method has not been reported.

The drug is administered by deep intramuscular injection as the oral administration is highly irritating and the intravenous route is dangerous because of cardiotoxicity. Human pharmacokinetic data of dehydroemetine are not available. The half-life of dehydroemetine in guinea pigs is about 2 days while that of emetine is 5 days (2, 3).

Clinical trials

In an uncontrolled study, dehydroemetine doses of 40–80 mg daily for 10 days cured 13 out of 18 patients (72%) with amoebic dysentery (4). A similar dose schedule and simultaneous treatment with a contact amoebicide (di-iodohydroxyquinoline) and/or chloroquine gave cure rates of 60–100% in small groups of patients (5). In a randomized clinical study in patients with amoebic liver abscess, intravenous metronidazole (1.5 g/day in 3 divided doses for 10 days, $n = 18$ patients) was compared to intramuscular dehydroemetine (60 mg/day for 10 days, $n = 18$ patients). After treatment, all patients treated with metronidazole were reported to be cured. However, 7 patients treated with dehydroemetine were switched to metronidazole, 2 because of toxicity and 5 because of failure of efficacy (6).

Indications

Serious intestinal amoebiasis and amoebic hepatitis with liver abscess. Dehydroemetine is usually given with chloroquine. It is only indicated when other safer drugs are not available or contraindicated.

Pregnancy and lactation

Documentation is lacking both in animals and in man. Because of its mechanism of action against protein synthesis and its severe side effects, the drug should not, however, be given during pregnancy, unless the condition of the patients makes its use necessary.

Its excretion into breast milk is unknown.

Side effects

Side effects are similar to those caused by emetine, but they are milder and less frequent. They are largely gastrointestinal, cardiac and neuromuscular reactions. Gastrointestinal reactions include nausea, vomiting, and abdominal pain. Neuromuscular effects are generalized muscle weakness, muscle pain and stiffness of the limbs and neck. Cardiac effects include fall in blood pressure, tachycardia and ECG changes of T-wave flattening or inversion (4, 5). The injection of dehydroemetine is painful and the formation of an abscess is common (4).

Contraindications and precautions

Patients with heart, muscle, or neurological diseases or in poor general condition should be treated with great caution. Dehydroemetine should always be given under medical supervision. Strenuous exercise should be forbidden until 4–5 weeks after completion of therapy (7). Dehydroemetine should not be given earlier than 1.5–2 months after emetine treatment because of cardiotoxicity.

Previous recommendations to use a contact amoebicide (e.g. di-iodohydroxyquinoline) together with an oral drug preparation can no longer be supported. The hydroxyquinolines have been withdrawn from most markets because of their potential risk of inducing subacute myelo-optic neuropathy (SMON).

Interactions

There have been no reports.

Dosage (8)

Adults

1 mg/kg daily, to a maximum of 60 mg, for up to 4–6 days. The dose should be reduced by up to 50% in the elderly and severely ill patients.

Children

1 mg/kg daily for no more than 5 days.

The drug should be given by deep intramuscular injection. At least 6 weeks should elapse before a second course is administered. In amoebic hepatitis with liver abscess, supplementary treatment with chloroquine is given orally. All patients should subsequently receive diloxanide by mouth to eliminate any surviving organisms in the colon.

Preparations

Available as dehydroemetine hydrochloride: 100 mg salt equals 87 mg base.

- Dehydroemetine® (Roche) Ampoules 30 mg/ml, ampoules 60 mg/2 ml.

References

1. Westwood JT, Wagenaar EB (1983). Chinese hamster cells can be reversibly blocked before mitosis with the protein synthesis inhibitor Emetine. *J Cell Sci*, **59**, 257–268.
2. Schwartz DE, Herrero J (1965). Comparative pharmacokinetic studies of dehydroemetine in quinea pigs using spectrophotometric methods. *Am J Trop Med Hyg*, **14**, 78–83.
3. Schwartz DE, Reider J (1961). Comparison of the rate of elimination of racemic 2-dehydroemetine (Ro 1-9334) and of natural emetine in animals. *Bull Soc Pathol Exot*, **54**, 38–48.
4. Salem HH, Abd-Rabbo H (1964). Clinical trials with dehydroemetine dihydrochloride in the treatment of acute amoebiasis. *J Trop Med Hyg*, **67**, 137–141.
5. Powell SW, Wilmot AJ, McLeod IN *et al.* (1965).Dehydroemetine in the treatment of amoebic liver abcess. *Ann Trop Med Parasitol*, **59**, 208–209.
6. Satpathy BK, Acharya SK, Satpathy S (1988). Comparative study of intravenous metronidazole and intramuscular dehydroemetine in Amoebic liver abscess. *J Indian Med Assoc*, **86**, 38–40.
7. Wilmot AJ (1962). *Clinical Amoebiasis* (Oxford: Blackwell Scientific Publications).
8. *WHO Model Prescribing Information. Drugs used in parasitic diseases* (1990), (Geneva: World Health Organization).

Diethylcarbamazine

Chemical structure

Diethylcarbamazine

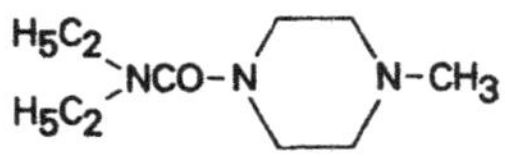

Physical properties

Base: MW 199; citrate: MW 391; pKa 7.7. Freely soluble in water. It should be stored in air-tight containers and the solutions protected from light.

Pharmacology and mechanism of action

Diethylcarbamazine (DEC) is a piperazine derivative which was introduced in clinical medicine in 1947. The drug is active against adult and microfilariae forms of *Wuchereria bancrofti*, *Brugia malayi*, *Brugia timori* and *Loa loa*. Against *Onchocerca volvulus*, the drug is only effective against the microfilariae. Soon after its administration, it causes rapid disappearance of microfilariae from the blood (lymphatic filariasis) or from the skin (onchocerciasis). Its effect against microfilariae in nodules or in hydroceles and in advanced elephantiasis is minimal (1).

The mechanism of action of DEC is not well understood. The drug has no microfilaricidal effect *in vitro* (2). *In vivo*, the microfilariae is first immobilized by the drug due to a possible hyperpolarization of the worm. This is followed by changes on the outer surface of the microfilariae making them more susceptible to the host's defence system (3). More recently, the drug has been reported to inhibit microtubule polymerization and disrupt preformed microtubule protein prepared from porcine brain *in vitro* (4). The relevance of this action in the living parasite remains to be studied. A similar mechanism of action has been described for benzimidazoles (see Mebendazole, p. 78).

Pharmacokinetics

Specific GC methods have been described for the determination of DEC (5, 6).

The drug is taken orally. Absolute bioavailability is not known, but urinary excretion data of a radiolabelled solution given to patients with onchocerciasis, indicated to be complete (>90%) absorption (7). After a single 50 mg tablet dose of DEC to healthy volunteers, peak plasma levels of 100–150 ng/ml were reached within 1 to 2 hours (7). The drug is distributed widely with an apparent volume of distribution of 240 l (range, 107 to 371 l) (7). The plasma protein binding of DEC is not known, but has been reported to be negligible (8). The drug is eliminated with a plasma half-life of 5 to 13 hours in both healthy volunteers and in patients with onchocerciasis (7).

Around 50% of the drug is recovered unchanged in the urine (7). However, in patients with chronic kidney impairment the renal excretion of the drug was substantially reduced (9). Approximately 10% of the drug is excreted as DEC-N-oxide with the urine, while a similar amount may be eliminated with the faeces (7).

DEC is a weakly basic compound and its urinary excretion is pH dependent. In one study, approximately 60% of the drug was excreted in the urine at pH 5, during the 48 hours following drug intake, while less than 10% of the same dose was excreted at pH 8 during a similar period (10). Alkalinization of urinary pH has not been successful in clinical practice because of increased toxicity without proportional increase of drug efficacy (11).

Clinical trials

In randomized, comparative double-blind studies, 200 mg of DEC given daily for 2–4 weeks against onchocerciasis was more effective in reducing skin microfilariae than mebendazole, 2 g/daily for 4 weeks or flubendazole, 750 mg i.m. once a week for 5 weeks (12, 13). In another study with a similar design, 50 mg of DEC given daily for 2 days followed by 100 mg twice daily for 6 days was compared with a single dose of 12 mg of ivermectin or a placebo (14). Microfilariae density decreased rapidly to 2% of pre-treatment values at day 8 in groups treated either with DEC or ivermectin. However, 12 months later, microfilariae densities rose up to 18% for the DEC treated group, while only 4% showed an increase of microfilariae density in the group treated with ivermectin. Ivermectin also caused fewer side effects than DEC (14). Several other well controlled studies comparing DEC with ivermectin have reported similar findings (15, 16). There is no evidence that transdermal DEC is microfilaricidal. Two double-blind trials have shown that this route of application is inefficient and causes an increased frequency of side effects (17, 18).

Large scale treatment programmes of lymphatic filariasis with diethylcarbamazine conducted in Malaysia, Haiti, India, Papua New Guinea, French Polynesia and in Kenya reported long term suppression of the filariaemia in those populations (19–22). A table salt medicated DEC was used in some of the studies.

Indications

DEC is used for the treatment of individual cases infected by *Wuchereria bancrofti*, *Brugia malayi*, *B. timori* and *Loa loa*. It may also be used for large scale chemotherapeutic control of filariasis. DEC is also used in *Acanthocheilonema streptocerca* infestations and in tropical eosinophilia.

In onchocerciasis, DEC should only be used when ivermectin is not available.

Pregnancy and lactation

Teratogenicity has not been reported in rats (23). Documentation in man is lacking. Because of its toxicity and possible abortifacient effect, (24) DEC should be avoided during pregnancy, unless there is strong indication for use.

Its excretion into breast milk is unknown.

Side effects

Side effects related to DEC are usually mild and include headache, general weakness, joint pains, anorexia, nausea and vomiting. They are dose-dependent.

There are specific side effects seen only in patients with filariasis. They are assumed to be caused by antigens released by dying microfilariae. In lymphatic filariasis, the side effects are usually mild. However, in onchocerciasis, the reaction to treatment with DEC may become quite severe. The reaction is known as the 'Mazzotti' after its original description and has been used as a diagnostic test for the disease (25). It occurs in two phases. A primary phase which commences within 24 hours and manifests as a variable combination of increased itching of the skin and eyes, photophobia, lacrimation, erythema and oedema of the skin and conjunctiva, lymphangitis, chills, anxiety, sweating and syncope. Respiratory distress, hyperpyrexia, hypotension, tachycardia and headache can also occur. Reversible proteinuria may be seen. A second phase may follow 2–6 days later with severe symmetrical acute polyarthritis predominantly in the knees, ankles, wrists, the interphalangeal joints and the shoulders. It is usually accompanied by a recrudescence of fever. The severity of Mazzotti reaction is related chiefly to the number of microfilariae killed (26).

To be able to quantify the severity of the reaction a scoring system has been developed (27). Using this scoring system, the suppressive effects of cyproheptadine, indomethacin, prednisolone, and their combinations on the reaction were evaluated. A marked suppression of the mean total reaction score occurred only in the group treated with the full course of prednisone.

Prednisone, however, had little effect on the severity of the itching and did not prevent the occurrence of the acute febrile polyarthritis of the secondary reaction. Prednisone has also significantly impaired the therapeutic effect of DEC (28, 29, 30).

Treatment may aggravate ocular lesions and precipitate blindness as a result of the reaction against the dead and dying microfilariae. A pre-treatment eye examination is advisable in cases with a high microfilarial density in a biopsy from epicanthus or from other locations.

Encephalitis and retinal damage may occur in patients with loaiasis. Periarticular swellings ('Calabar swelling') due to a local reaction around the dying worm are also frequently seen in such patients.

Contraindications and precautions

There are no known contraindications to the drug. Dosage should be reduced in patients with renal impairment (9) or in strict vegetarians with high urinary pH (10) since renal function and pH are important factors for the excretion of the drug. Dosage may also be reduced in patients in poor general condition or who are heavily infected.

Dosage (31)

Dosage is expressed in mg base (100 mg citrate is equivalent to 50 mg base).

Loa Loa

Treatment

Adults

Day 1: 1 mg/kg as a single dose.
Day 2: 2 mg/kg as a single dose.
Day 3: 4 mg/kg as a single dose.
Day 4–18: 2–3 mg/kg three times daily.

Prophylaxis:
Adults
300 mg once weekly for as long as exposure continues.

Wuchereria bancrofti

Individual treatment
6 mg/kg daily for 12 days administered orally, preferably in divided doses after meals.

Mass treatment
6 mg/kg as a single oral dose at weekly or monthly intervals or as a single annual dose.

Brugia malayi and B. timori

Individual treatment
3–6 mg/kg daily for 6–12 days administered orally, preferably in divided doses after meals.

Mass treatment
3–6 mg/kg as a single oral dose given 6 times at weekly or monthly intervals.

Several trials have shown that, used consistently over a period of at least 6 months, table salt medicated with DEC at a concentration of 0.1% can eliminate *W. bancrofti*. A concentration of 0.3% for 3–4 months may be necessary in areas where *B. malayi* is endemic.

Onchocerciasis

In onchocerciasis, the treatment of choice is presently ivermectin. However, because of some restrictions on its distribution, DEC may still be used in some areas. The dosage regimen of DEC varies widely and is largely based on trial and error. To reduce the acute reactions related to the use of the drug, earlier studies recommended that treatment start with small doses and that they should gradually increase (32). However, recent studies have reported little advantage in this prolonged treatment regimen. It is argued that since side effects occur even with small doses, prolonging the treatment will only expose the patient to longer suffering and eventually result in poor patient compliance. Fulford *et al.* (33) have recently estimated, the optimal dosage regimen of DEC in onchocerciasis after using a dose-response curve of pooled data from 10 studies conducted in Ghana during 1978 until 1983 and comprising 401 patients. The study concluded several important points:

1. A total dose of 2000 mg of DEC can reduce the microfilarial density by more than 96%, increasing the dose beyond this has little additional effect.
2. Only minor increase in microfilaricidal efficacy occurs when the total dose is increased from 1300 mg to 2000 mg.
3. The ED90 is approximately 500–600 mg.
4. Total reaction increases with total doses, the reaction to a dose of 6600 mg being significantly greater than to the 2000 mg dose although the reduction achieved in skin microfilarial counts was similar.
5. With a total dose exceeding 1000 mg, skin microfilariae counts will not increase above 40 microfilariae per 4 mg of skin for approximately 4 months (33).

Based on these findings the authors recommend a total dose of 1350 mg of DEC given over an 8-day period which is repeated every 4 months. A recommended dose for a 50 kg adult is:

Day 1: 50 mg.
Day 2: 50 mg twice.
Day 3–8: 100 mg twice daily.

Experience with this regimen, however, is still limited.

The dosage regimen recommended by the WHO (31) is as follows:

Adults
Day 1 0.5 mg/kg.
Day 2: 0.5 mg/kg twice daily.
Day 3: 1 mg/kg twice daily.
Day 4–9: 4–5 mg/kg divided in two daily doses.

Children
Initially 1 mg/kg should be given on 2 successive days. This is then raised incrementally, firstly to 2 mg/kg daily, and subject to tolerance, to 4 mg/kg daily. Any adverse effect should be allowed to subside before the subsequent dose is administered. The full daily dose of 4 mg/kg is usually attained within 7–14 days. This should be continued for 2 further weeks.

In both adults and children, suramin is given subsequently during 5 weeks to kill the adult filariae. Thereafter, a repeated treatment course of diethylcarbamazine is given to kill microfilariae derived from adult filariae surviving the suramin treatment. In patients with a low density of microfilariae and without eye involvement, a higher dose schedule may be used. In these patients it is not advisable to give suramin because of its toxicity.

Tropical eosinophilia

2 mg/kg three times daily for 7–10 days.

Preparations

Available as diethylcarbamazine citrate: 100 mg citrate is approximately equivalent to 50 mg base.

- Banocide® (Wellcome) Oral solution 10 mg/ml and 24 mg/ml; tablets 50 mg, 100 mg.
- Hetrazan® (Lederle) Tablets 50 mg.
- Notezine® (Specia) Tablets 50 mg.

References

1. Webster LT (1990). Chemotherapy of parasitic infections. In: *Goodman & Gilman's The Pharmacological basis of Therapeutics*, 8th edn, edited by A. G. Gilman, T. W. Rall, A. S. Nies and P. Taylor (New York: Pergamon Press) pp. 960–961.
2. Langham ME, Kramer TR (1980). The *in vitro* effect of diethylcarbamazine on the motility and survival of *Onchocerca volvulus* microfilariae. *Tropenmed Parasitol*, **31**, 155–158.
3. Hawking F (1979). Diethylcarbamazine and new compounds for the treatment of filariasis. *Adv Pharmacol Chemother*, **16**, 129–194.
4. Fujimaki Y, Ehara M, Kimura E, Shimada M, Aoki Y (1990). Diethylcarbamazine, antifilarial drug, inhibits microtubule polymerisation and disrupts preformed microtubules. *Biochem Pharmacol*, **39**, 851–856.
5. Allen GD, Goodchild TM, Weatherley BC (1979). Determination of 1-diethylcarbamoyl-4-methylpiperazine in human plasma and urine. *J Chromatogr*, **164**, 521–526.
6. Nene S, Anjaneyulu B, Rajagopolan TG (1984). Determination of diethylcarbamazine in blood using gas chromatography with alkali flame ionization detection. *J Chromatogr*, **308**, 334–340.

7. Edwards G, Awadzi K, Breckenridge AM, Gilles HM, L'E Orme M, Ward SA (1981). Diethylcarbamazine disposition in patients with onchocerciasis. *Clin Pharmacol Ther*, **30**, 551–557.
8. Edwards G, Breckenridge A (1988). Clinical pharmacokinetics of anthelminthic drugs. *Clin Pharmacokinet*, **15**, 67–93.
9. Adjepon-Yamoah KK, Edwards G, Breckenridge AM, Orme ML'E, Ward SA (1982). The effect of renal disease on the pharmacokinetics of diethylcarbamazine in man. *Br J Clin Pharmacol*, **13**, 829–834.
10. Edwards G, Breckenridge AM, Adjepon-Yamoah KK, Orme ML'E, Ward SA (1981). The effect of variations in urinary pH on the pharmacokinetics of diethylcarbamazine. *Br J Clin Pharmacol*, **12**, 807–812.
11. Awadzi K, Adjepon-Yamoah KK, Edwards G, Orme ML'E, Breckenridge AM, Gilles HM (1986). The effect of moderate urine alkalinization on low dose diethylcarbamazine therapy in patients with onchocerciasis. *Br J Clin Pharmacol*, **21**, 669–676.
12. Rivas-Alcalá AR, Greene BM, Taylor HR, Domiquez-Vázquez A, Ruvalcaba-Macías AM, Lugo-Pfeiffer C, Mackenzie CD, Beltrán F (1981). Chemotherapy of onchocerciasis: a controlled comparison of mebendazole, flubendazole and diethylcarbamazine. *Lancet*, **II**, 485–490.
13. Domiquez-Vázquez A, Taylor HR, Greene BM, Ruvalcaba-Macías AM, Rivas-Alcalá AR, Murphy RP, Beltran-Hernandez F (1983). Comparison of flubendazole and diethylcarbamazine in treatment of onchocerciasis. *Lancet*, **I**, 39–143.
14. Diallo S, Aziz MA, Larivière M, Diallo JS, Diop-Mar I, N'dir O, Badiane S, Py D, Schulz-Key H, Gaxotte P, Victorius A (1986). A double blind comparison of the efficacy and safety of ivermectin and diethylcarbamazine in a placebo controlled study of Senegalese patients with onchocerciasis. *Trans R Soc Trop Med Hyg*, **80**, 927–934.
15. Taylor HR, Murphy RP, Newland HS, White AT, D'Anna SA, Keyvan-Larijani E, Aziz MA, Cupp EW, Greene BM (1986). Treatment of onchocerciasis. The ocular effects of ivermectin and diethylcarbamazine. *Arch Ophtalmol*, **104**, 863–870.
16. Albiez EJ, Newland HS, White AT, Kaiser A, Greene BM, Taylor HR, Büttner DW (1988). Chemotherapy of onchocerciasis with high doses of diethylcarbamazine or a single dose of ivermectin: microfilariae levels and side effects. *Trop Med Parasitol*, **39**, 19–24.
17. Taylor HR, Greene BM, Langham ME (1980). Controlled clinical trial of oral and topical diethylcarbamazine in the treatment of onchocerciasis. *Lancet*, **I**, 943–946.
18. Taylor HR, Langham ME, de Stahl EM, Figueroa LN, Beltranena F (1980). Chemotherapy of onchocerciasis: a controlled clinical trial of topical diethylcarbamazine (DEC) in Guatemala. *Tropenmed Parasitol*, **31**, 357–364.
19. Wijers JDB, Kaleli N, Ngindu AH (1988). Diethylcarbamazine prophylaxis against bancroftian filariasis given by a member of the local community in Kenya. *Ann Trop Med Parasitol*, **82**, 411–412.
20. Lu King HII J, Kin Ping Kan S, Parmar SS, Kin Chung Chan M, Wah Mak J, Kim Chool Lim P, Wah Lim T, Dennis DT (1988). The effect of diethylcarbamazine citrate on incidence and recovery rates of *Brugia malayi* microfilaremia in Sabah, Malaysia. *Am J Trop Med Hyg*, **38**, 582–588.
21. Fan PC (1990a). Eradication of bancroftian filariasis by diethylcarbamazine-medicated common salt on little Kinmen (Liehyu district), Kinmen (Quemoy) Island, Republic of China. *Ann Trop Med Parasitol*, **84**, 25–33.
22. Jingyuan L, Zi C, Xiaohang H, Zhaoping T (1992). Mass treatment of filariasis using DEC-medicated salt. *J Trop Med Hyg*, **95**, 132–135.
23. Diethylcarbamazine. *Therapeutic Drugs*, edited by Sir Colin Dollery (1991), (London: Churchill Livingstone), pp. D109-D112.
24. Joseph CA, Dixon PAF (1984). Possible prostaglandin-mediated effect of diethylcarbamazine on rat uterine contractility. *J Pharm Pharmacol*, **36**, 281–282.
25. Mazzotti L (1948). Posibilidad de utilizar como medio diagnóstico auxiliar en la oncocercosis las reacciones alérgicas consecutivas a la administración del 'Heterzán'. *Revista del Instituto Salubridad Enfermedades Tropicales*, **9**, 235–237.
26. Awadzi K, Gilles HM (1992). Diethylcarbamazine in the treatment of patients with onchocerciasis. *Br J Clin Pharmacol*, **34**, 281–288.
27. Awadazi K (1980). The chemotherapy of onchocerciasis. II. Quantification of the clinical reaction to microfilaricides. *Ann Trop Med Parasitol*, **74**, 189–197.
28. Awadzi K, Orme ML'E, Breckenridge AM, Gilles HM (1982). The chemotherapy of onchocerciasis VI. The effects of indomethacin and cyproheptadin on the Mazzotti reaction. *Ann Trop Med Parasitol*, **76**, 323–330.

29. Awadzi K, Orme ML'E, Breckenridge AM, Gilles HM (1982). The chemotherapy of onchocerciasis VII. The effect of prednisone on the Mazzotti reaction. *Ann Trop Med Parasitol*, **76**, 331–338.
30. Awadzi K, Orme ML'E, Breckenridge AM, Gilles HM (1982). The chemotherapy of onchocerciasis IX: The effect of prednisone plus cyproheptadine on the Mazzotti reaction. *Ann Trop Med Parasitol*, **76**, 547–555.
31. *WHO Model Prescribing Information. Drugs used in parasitic diseases*. (Geneva: World Health Organization).
32. Taylor HR, Greene BM, Langham ME (1980). Controlled clinical trial of oral and topical diethylcarbamazine in treatment of onchocerciasis. *Lancet*, **II**, 943–946.
33. Fulford AJ, Macfarlane SB, Awadzi K, Bell DR, Gilles HM (1987). The chemotherapy of onchocerciasis. XII. The prediction of microfilarial loads in patients with onchocerciasis after treatment with diethylcarbamazine in northern Ghana. *Ann Trop Med Parasitol*, **78**, 701–711.

Diloxanide

Chemical structure

Diloxanide

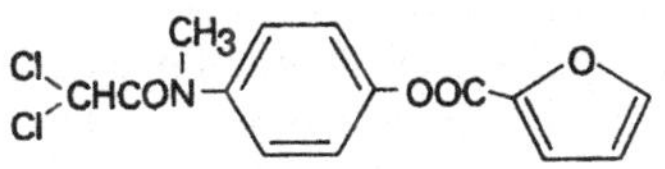

Physical properties

Diloxanide furoate: MW 328. Neutral compound. Almost insoluble in water. Protect from light.

Pharmacology and mechanism of action

Diloxanide is a dichloroacetanilide derivative that was introduced in 1956. It is amoebicidal *in vivo* and *in vitro*. It is highly effective in asymptomatic patients passing cyst forms. Sufficient data are not available on its efficacy when used alone in acute amoebiasis (1).

The mechanism of action of diloxanide is unknown. Like the structurally related chloramphenicol, diloxanide has been suggested to block the protein synthesis in the microorganism (2).

Pharmacokinetics

Specific analytical methods have not been reported.

The drug is only given orally. Human pharmacokinetic data are lacking. According to animal studies, the drug is slowly absorbed, resulting in high intraluminal concentrations in the gut. The ester is hydrolysed in the intestine to diloxanide and furoic acid. The amount absorbed is excreted primarily through the kidneys as the glucuronide during the first 6 hours, but excretion may continue up to 48 hours. Less than 10% of the dose is eliminated with the faeces (3).

Clinical trials

Diloxanide gives a cure rate above 90% when treating non-invasive forms of amoebiasis in patients with no risk of re-infection using standard doses for 10 days (4, 5, 6). In a retrospective study, the efficacy of diloxanide given to 1,535 asymptomatic patients during 1984–1990 was evaluated. Of 539 patients with complete follow-up, 497 (86%) were reported as cured (7). Lower cure rates have been reported in trials performed in tropical countries where patients are exposed to reinfection during the follow-up period (8, 9).

Indications

Diloxanide is the drug of choice in the treatment of asymptomatic passers of cysts of *Entamoeba histolytica* in non-endemic countries. It is also given after metronidazole treatment to eradicate residual amoeba in the intestine.

Pregnancy and lactation

Diloxanide has not been shown to be teratogenic in rats and rabbits (10). Documentation in man is lacking. Diloxanide is probably safe in pregnancy, but its indication may not justify use during the first trimester.

Its excretion into breast milk is unknown.

Side effects

Diloxanide is usually well tolerated even at high doses. In one study (5), excessive flatulence was the only significant side effect recorded in 87% of the patients, but was also a common complaint among the patients (31%) even before treatment. Other minor side effects included anorexia (3%), nausea (6%), diarrhoea (2%), and abdominal cramps (2%). Flatulence as a frequent side effect of diloxanide has also been reported in a large retrospective study covering more than 4000 patients who used the drug during 1977 until 1990 in the United States (7). Other less frequent side effects reported in the survey included headache, lethargy, dizziness, diplopia, and paraesthesia. The percentage of persons reporting adverse effects varied significantly by racial group. The existence of racial differences in the metabolism of diloxanide is unknown.

Contraindications and precautions

There are no known contraindications to the drug.

Interactions

There have been no reports.

Dosage

The dose is expressed in mg of the diloxanide furoate.

Adults
0.5 g 3 times daily for 10 days.
Children
20 mg/kg daily divided into 3 doses for 10 days.

Preparations

Available as diloxanide furoate.

- Furamide® (Boots) Tablets 500 mg.

References

1. Krogstad DJ, Spencer HC Jr, Healy GR (1978). Amoebiasis. *N Engl J Med*, **298**, 262–265.
2. Knight R (1980). The chemotherapy of amoebiasis. *J Antimicrob Chemother*, **6**, 557–593.
3. Wilmshurst EC, Cliffe EE (1964). Absorption and distribution of amoebicides. In: *Absorption and Distribution of Drugs*, edited by T. B. Binns (Edinburgh: E. S. Livingstone), p. 191.
4. Woodruff HW, Bell S (1960). Clinical trials with entamide furoate and related compounds. I: in a non-tropical environment. *Trans R Soc Trop Med Hyg*, **54**, 389–395.
5. Wolfe MS (1973). Non-dysenteric intestinal amoebiasis: Treatment with diloxanide furoate. *J Am Med Ass*, **224**, 1601–1604.
6. Thoren K, Håkansson C, Bergström T, Johansson G, Norkrans G (1990). Treatment of

asymptomatic amoebiasis in homosexual men. Clinical trials with metronidazole, tinidazole and diloxanide furoate. *Sex Transm Dis*, **17**, 72–74.

7. McAuley JB, Herwaldt BL, Stokes SL, Becher JA, Roberts JM, Michelson MK, Juranek DD (1992). Diloxanide furoate for treating asymptomatic *Entamoeba histolytica* cyst passers: 14 years' experience in the United States. *Clin Infect Dis*, **15**, 464–468.
8. Bell S (1967). An investigation of carriers of *Entamoeba histolytica*. *Trans R Soc Trop Med Hyg*, **61**, 506–513.
9. Forsyth DM (1967). The treatment of amoebiasis: a field study of various methods. *Trans R Soc Trop Med Hyg*, **61**, 506–513.
10. Diloxanide. *Therapeutic Drugs*. edited by Sir Colin Dollery (1991), (London: Churchill Livingstone) pp. D140–D143.

Eflornithine

Chemical structure

Eflornithine

$$H_2NCH_2CH_2CH_2\overset{\displaystyle CHF_2}{\underset{\displaystyle NH_2}{C}} - COOH$$

Physical properties

Base: MW 182; hydrochloride: MW 237; pKa not known.

Pharmacology and mechanism of action

Eflornithine, formerly known as DFMO (α-difluoromethylornithine), is a recent product deliberately designed to inhibit polyamine synthesis. The drug was originally intended for tumour chemotherapy but was later found unsatisfactory. The drug has demonstrated antiprotozoal activity *in vitro*, particularly against *Trypanosoma brucei gambiense* (1) and *Pneumocystis carinii* (2). The efficacy of the drug in human trypanosomes has been confirmed both in animal models (3) and in humans (4). In 1990 the US Food and Drug Administration approved eflornithine for the treatment of *Trypanosoma brucei gambiense*.

The mechanism of action of eflornithine is due to its irreversible inhibition of ornithine decarboxylase (ODC) which catalyses the biosynthesis of polyamines. Common polyamines such as putrescine, spermidine and spermine are low-molecular weight molecules present in all living cells. They are important for cell growth, differentiation and replication of trypanosomes. Trypanosomes are more susceptible to the drug than human cells, possibly due to their slow turnover of this enzyme (1).

Pharmacokinetics

Specific HPLC methods have been described for the determination of eflornithine (5, 6).

The drug is given orally or intravenously. Absolute oral bioavailability is around 50% (7). In 6 healthy volunteers given single oral doses of 10 and 20 mg/kg of eflornithine hydrochloride, peak plasma levels of 39 and 77 μmol/l, respectively, were obtained 4 hours after drug administration. With this dosage range, the amount of drug absorbed was directly proportional to the dose given. However, when higher doses were used, non-linear absorption kinetics were observed with no increase in area under the curves (8). The drug is quickly distributed with a mean apparent volume of distribution of 0.34 l/kg (7). It is apparently not metabolized in the body. It is quickly eliminated with a mean plasma elimination half-life of 3 hours. More than 80% of the drug was recovered unchanged in the urine during the first 24 hours after drug intake (7).

Eflornithine passes into the CSF. In a recent study with 63 patients with trypanosomiasis, the mean CSF/plasma ratio of eflornithine measured at the end of a 14 day course of intravenous administration was 0.91 in adults and 0.58 in children less than 12 years (9).

Clinical trials

The available clinical data on eflornithine for the treatment of trypanosomiasis are based on sporadic case reports and uncontrolled clinical observations from Sudan, Italy, Belgium, Côte d'Ivoire, USA, Republic of Zaire, France, and the Peoples Republic of Congo (4, 10–16). The results of 711 patients treated from 1981 until 1990 has recently been reviewed by Hardenberg *et al.* (17). Of these, 675 (95%) were in the late stage, while 36 (5%) were in the early stage of the disease. The dosage regimens were as follows: 252 patients received 100 mg/kg intravenously every 6 hours for 14 days followed by 75 mg/kg orally every 6 hours for 21–30 days; 324 patients received 100–200 mg/kg intravenously every 6–12 hours for 14 days; and 135 patients received 100 mg/kg orally every 6 hours for 21–45 days. After treatment, 27% were cured (follow-up at least 24 months), while 22% were probably cured (follow-up at least 12 months). Of 348 patients followed for up to 12 months, a higher relapse rate was reported in those under 12 years old compared with grown-ups (36% *vs* 7%). Relapses were also more common in those patients who received the drug orally compared with those who received it intravenously. Side effects were common but reversible after stopping treatment. In all, 49 (7%) patients died during or shortly after treatment.

In a more recent open clinical trial, Milord *et al.* (18) treated 207 patients with late-stage *Trypanosoma brucei gambiense* sleeping sickness in Zaire. The patients were treated with 3 different dosage regimens of 100 mg/kg i.v. every 6 hours for 14 days, followed by 75 mg/kg i.v. every 6 hours for 21 days; or 200 mg/kg i.v. every 12 hours for 14 days; or 75 mg/kg p.o. every 6 hours for 35 days. Of 152 patients followed for at least a year, only 9% relapsed. Treatment failures were more common in children less than 12 years old and patients who received oral treatment. Side effects were common and 4 patients died during or shortly after treatment.

Indications

Because of its high cost, need of repeated intravenous administration and the relatively large quantities of the drug needed for each patient eflornithine use will be associated with major difficulties in rural Africa. Currently it is only recommended in patients with late-stage *Trypanosoma brucei gambiense* sleeping sickness refractory to melarsoprol.

Pregnancy and lactation

Teratogenicity has not been reported in animals. However, the drug arrests embryonic development in mice, rats and rabbits (19, 20). Documentation in man is lacking. Eflornithine should not be given during pregnancy, unless there is a strong indication for use.

Its excretion into breast milk is unknown.

Side effects

Eflornithine is generally better tolerated than melarsoprol. In one study in Zaire (18), where 207 patients were given the drug the following side effects were reported: leucopenia (53%), anaemia (43%), diarrhoea (13%), convulsions (4%), abdominal pain (3%). Four patients (2%), who were already in a severe condition before treatment, died during treatment. Diarrhoea accompanied with abdominal pain was usually encountered after oral administration of the drug. Anaemia developed several weeks after treatment.

Other side effects reported include alopecia, hearing loss, blood in stool, thrombocytopenia, haematuria and alterations in liver function and skin rashes (10, 18). All side effects are reported to be reversible after drug discontinuation.

Contraindications and precautions

Patients with cardiac diseases, epilepsy or with anaemia must be treated with extra caution. White blood cell counts and haemoglobin levels must be monitored during eflornithine treatment. In patients with kidney failure dosage reduction has to be made.

Interactions

Drugs such as melarsoprol, suramin, and antimonial compounds have been reported to potentiate the clinical effects of eflornithine (21–23). Combination between melarsoprol and eflornithine seems rational since both drugs have effects on trypanathione.

Dosage

The dosage regimen of eflornithine is complicated. In most areas, a dosage regimen of 100 mg/kg of eflornithine hydrochloride i.v. every 6 hours for 14 days, followed by 75 mg/kg orally every 6 hours for 21 days have been used. However, the current opinion is that nothing is gained by the addition of the 3 more weeks of oral therapy (18).

Preparations

Available as eflornithine hydrochloride.

- Ornidyl® (Marion Merrel Dow) Solution for injection, 100 mg per ml.

Ornidyl is not available in all countries. The drug may be obtained from the World Health Organization, Geneva, Switzerland (attention: Dr Kuzoe).

References

1. Bacchi CJ, Nathan HC, Hunter SH (1980). Polyamine metabolism: a potential therapeutic target in trypanosomes. *Science*, **210**, 332–334.
2. Cushion MT, Stanforth D, Linke MJ, Walzer PD (1985). Method of testing the susceptibility of *Pneumocystis carinii* to antimicrobial agents *in vitro*. *Antimicrob Agents Chemother*, **28**, 796–801.
3. McCann PP, Bachi CJ, Clarkson AB Jr, Seed JR, Nathan HC, Amole BO, Hutner SH, Sjoerdsma A (1981). Further studies on difluoromethylornithine in African trypanosomes. *Med Biol*, **59**, 434–440.
4. Van Nieuwenhove S, Schechter PJ, Declercq J, Burke J, Sjoerdsma A (1985). Treatment of *gambiense* sleeping sickness in the Sudan with oral DFMO (DL-alfa-difluoromethylornithine) an inhibitor of ornithine decarboxylase; first field trial. *Trans R Soc Trop Med Hyg*, **79**, 692–698.
5. Cohen JL, Ko RJ, Lo ATL, Shields MD, Gilman TM (1989). High pressure liquid chromatographic analysis of eflornithine in serum. *J Pharm Sci*, **78**, 114–116.
6. Smithers J (1988). A precolumn derivatization high performance liquid chromatographic (HPLC) procedure for the quantitation of difluoromethylornithine in plasma. *Pharm Res*, **5**, 684–686.
7. Haegele KD, Alken RG, Grove J, Schechter PJ, Koch-Weser J (1981). Kinetics of alpha-difluoromethylornithine: An irreversible inhibitor of ornithine decarboxylase. *Clin Pharmacol Ther*, **30**, 210–217.

8. Griffin CA, Slavik M, Chien SC, Hermann J, Blanc O, Luk GD, Baylin SD (1987). Phase 1 trial and pharmacokinetics study of intravenous and oral alpha-difluoromethylornithine. *Investigational New Drugs*, **5**, 177–186.
9. Milord F, Loko L, Ethier L, Mpia B, Pepin J (1993). Eflornithine concentrations in serum and cerebrospinal fluid of 63 patients treated for *Trypanosoma brucei gambiense* sleeping sickness. *Trans R Soc Trop Med Hyg*, **87**, 473–477.
10. Doua F, Boa FY, Schechter PJ, Miezan TW, Haegele KD, Sjoerdsma A, Konian K (1987). Treatment of human late stage *gambiense* trypanosomiasis with alfa-difluoromethylornithine (eflornithine): Efficacy and tolerance in 14 cases in Côte D'Ivoire. *Am J Trop Med Hyg*, **37**, 525–533.
11. Taelman H, Schechter PJ, Marcelis L, Sonnet J, Kazyumba G, Dasnoy J, Haegele K, Sjoerdsma A, Wery M (1987). Difluoromethylornithine, an effective new treatment of Gambian trypanosomiasis – Results in five patients. *Am J Med*, **82**, 607–614.
12. Eozenou P, Jannin J, Ngampo S, Carme B, Tell GP, Schechter PJ (1989). Essai de traitement de la trypanosomiase à *Trypanosoma brucei-gambiense* par l'éflornithine.en République populaire du Congo. *Med Trop*, **49**, 149–154.
13. Petru A. M, Azimi PH, Cummins SK, Sjoerdsma A (1988). African sleeping sickness in the United States: Successful treatment with eflornithine. *Am J Dis Child*, **142**, 224–228.
14. Pepin J, Guern C, Milord F, Ethier Bokelo M, Schechter PJ (1989). Utilisation de la difluorométhylornithine dans la trypanosomiase congénitale à *Trypanosoma brucei gambiense*. *Med Trop*, **49**, 84–85.
15. Benhamou PH, Chandenier J, Schechter PJ, Epelbaum S, Tell GP, Haegele KD, Pautard JC, Piussan CH (1989). Trypanosomiase africaine de l'enfant traité par éflornithine (case report). *Presse Méd*, **24**, 1199–1202.
16. Sjoerdsma A, Golden JF, Schechter PJ, Barlow JLR, Santi D (1984). Successful treatment of lethal protozoal infections with the ornithine decarboxylase inhibitors alpha-difluoromethylornithine. *Trans Assoc Am Physicians*, **97**, 70–79.
17. Hardenberg J, Claverie N, Tell GP (1991). Eflornithine (Ornidyl) treatment of *Trypanosoma brucei gambiense* sleeping sickness; report of 711 patients treated up to March 1991. Presented at the 21st meeting of International Scientific Council for Trypanosomiasis Research and Control, (Yamoussoukro, Côte d'Ivoire, October); abstr 412B.
18. Milord F, Pepin J, Loko L. Mpia B (1992). Efficacy and toxicity of eflornithine for treatment of *Trypanosoma brucei gambiense* sleeping sickness. *Lancet*, **340**, 652–655.
19. Fozard JR, Part ML, Parakash NJ, Grove J, Schechter PJ, Sjoerdsma A, Koch-Weser J (1980). L-Ornithine decarboxylase: An essential role in early mammalian embryogenesis. *Science*, **208**, 505–508.
20. O'Toole BA, Huffman KW, Gibson JP (1989). Effects of eflornithine hydrochloride (DFMO). *Teratology* **39**, 103–113.
21. Jennings FW (1988). Chemotherapy of trypanosomiasis: the potentiation of melarsoprol by concurrent difluoromethylornithine (DFMO) treatment. *Trans R Soc Trop Med Hyg*, **82**, 572–573.
22. Jennings FW (1991). Chemotherapy of trypanosomiasis: the potentiation of antimonial compounds by difluoromethylornithine (DFMO). *Trop Med Parasitol*, **42**, 135–138.
23. Clarkson AB, Bienen EJ, Bacchi CJ, McCann PP, Nathan HC, Hunter SH, Sjoerdsma A (1984). New drug combination for experimental late-stage African trypanosomiasis: DL-alfa-difluoromethylornithine (DFMO) with suramin. *Am J Trop Med Hyg*, **33**, 1073–1077.

Halofantrine

Chemical structure

Halofantrine

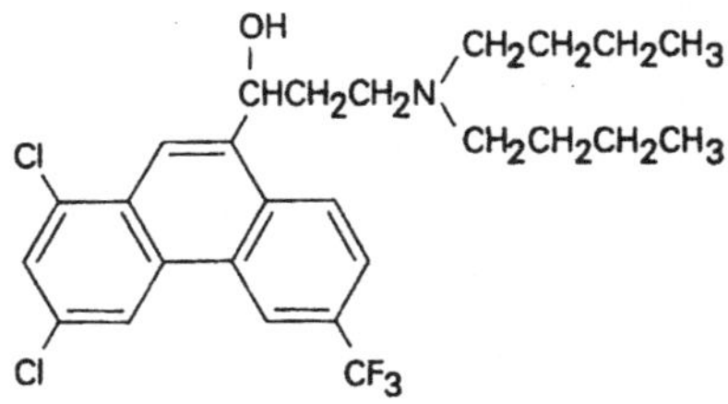

Physical properties

Base: MW 500; hydrochloride MW 537; pKa 9.7. Less than 1 g dissolves in 100 ml of water.

Pharmacology and mechanism of action

Halofantrine is a phenanthrenemethanol antimalarial drug developed by the US military. The antimalarial activity of the phenanthrenemethanols was discovered during the Second World War but was not exploited until later. Halofantrine is a potent blood schizontocide against *Plasmodium falciparum* both *in vitro* and *in vivo*. However, it has no effect against exoerythrocytic forms of the parasite. Experience of its activity against other malaria species is limited. The drug exists as a racemic mixture, but the two enantiomers have shown similar activity *in vitro* (1, 2).

The mechanism of action of halofantrine is not known.

Pharmacokinetics

Several specific and sensitive HPLC methods have been described for halofantrine and its active metabolite N-desbutylhalofantrine. Comparing their simplicity, sample volume used, sample pre-treatment procedure, and run-time, the methods of Keeraithakul *et al.* and Mberu *et al.* can be recommended for pharmacokinetic studies (3, 4).

Halofantrine is administered orally. It is poorly and erratically absorbed and large intra- and inter-individual variability in the plasma concentrations of the drug has been observed (5). Peak plasma levels of halofantrine varied form 65 to 392 μg/l 6 hours after a single 250 mg dose to healthy volunteers in a fasting state. The amount of drug absorbed seemed to increase disproportionately with doses above 500 mg. A three-fold increase in the bioavailability was found when the drug was taken with a fatty meal (5).

After oral administration, a wide range of mean terminal half-lives from 10.5 to 158 hours has been reported (6). In general, the pharmacokinetic properties of halofantrine have been difficult to assess due to poor and variable absorption.

The development of a new parenteral formulation of halofantrine has made detailed pharmacokinetic assessment possible. In a recent study 1 mg/kg was given as an intravenous infusion over 1 hour on three occasions with 8 hour intervals to Thai patients with *falciparum* malaria ($n = 12$) or in convalescence period ($n = 9$) (7). The pharmacokinetic data were fitted to a two-compartment open model. At the time of acute malaria, the initial half-life was 0.19 hours followed by a terminal half-life of 14.4 hours. The volume of distribution at steady state was 4.8 l/kg. In the convalescent period the terminal half-life was reduced to 7.5 hours (7). A prolonged half-life during malaria compared to in healthy volunteers has also been reported by others (8, 9).

Halofantrine is extensively metabolized to N-desbutylhalofantrine which has antimalarial activity (6). The plasma elimination half-life of the metabolite in healthy volunteers was about 11 days (9).

The route of excretion of halofantrine and its metabolite in man has yet to be determined. In animals (rats, dogs and monkeys) most of the drug is eliminated via the faeces indicating poor absorption or biliary excretion (6). No halofantrine was detected in the urine of subjects given the drug. Only insignificant amounts (<0.01%) of N-desbutylhalofantrine were recovered (5).

Clinical trials

Clinical studies have been reported from Rwanda (10), Kenya (11), Nigeria (12), Thailand (13), Gabon (14), and Malawi (15). These have established that a single dose of 500 mg (8 mg/kg) repeated thrice at 6-hourly intervals is safe and effective against chloroquine-resistant as well as chloroquine-sensitive *P. falciparum* parasites for both adults and children with cure rates of between 80% and 100%. When halofantrine was given as a single dose or a 2-dose regimen, numerous recurrences of the disease were noted (16). In non-immune, mostly German, patients with *falciparum* malaria, the cure rate increased from 85% to 100% with an additional treatment course (3 × 500 mg) on day 7 (17).

In a small study from Rwanda (10), 3 doses of halofantrine (500 mg every 6 hours) were shown to be effective in *P. falciparum* infections with high parasitaemia. In a similar study in Thailand, halofantrine (3 doses of 500 mg 6-hourly) was compared to mefloquine (25 mg/kg) in a multidrug resistant area on the Thai–Burmese border. The cumulative failure rates by day 28 were 35% with halofantrine and 10% with mefloquine. When the doses of halofantrine were increased to a total of 72 mg/kg, halofantrine was more effective and better tolerated than mefloquine (18). However, this high dose of halofantrine has been associated with significant cardiotoxicity (19). Although most of the treatment failures of halofantrine have been associated with poor absorption of the drug (6), a marked reduction of its efficacy due to development of parasite resistance was observed in a study conducted at the Thai–Cambodian border. Cure rates of the drug declined from 90% to 30% over 5 years (20).

Indications

Halofantrine is indicated only for the treatment of multidrug-resistant *Plasmodium falciparum* malaria.

Pregnancy and lactation

Halofantrine is not teratogenic in rats and rabbits. It was, however, found to be embryotoxic at doses of 30 mg base/kg per day in rats and 60 mg base/kg per day in rabbits (6).

Documentation in man is lacking. Halofantrine should not be given during pregnancy, unless there is a strong indication for use.

Its excretion into the breast milk is not known.

Side effects

Halofantrine is generally well tolerated. Mild and transient side effects such as nausea, vomiting, diarrhoea, abdominal pain, pruritus and rash have been reported in humans. Halofantrine is potentially cardiotoxic particularly with doses above the recommended dose and causes ECG changes such as prolongation of PR and QTc intervals. In one study, the sudden death of a patient was reported after receiving a high dose of halofantrine (8 mg/kg 3 times daily for 3 days) (19). The patient had previously been treated with mefloquine. Cardiotoxicity due to halofantrine will become a therapeutic problem if higher dosage regimens have to be used due to decreased efficacy (19). Occasional elevation of serum transaminase have been observed in some patients. The relationship of this to the treatment is unclear. Values usually return to normal levels within a week after treatment (11, 12, 13).

Contraindications and precautions

Halofantrine should not be given to patients with pre-existing cardiovascular diseases. There is a warning against the concomitant intake of any cardiotoxic drugs. Halofantrine is not used for malaria prophylaxis.

Interactions

There are no reports of interactions (6).

Dosage (1)

Adults and children (>40 kg)
3 doses of 500 mg (8 mg/kg) every 6 hours. For non-immune patients a second course of treatment is recommended after 7 days.
Children (<40 kg)
Similar to above (8 mg/kg) but with halofantrine suspension.

Preparations

Available as halofantrine hydrochloride: 100 mg hydrochloride is equal to 93 mg base. Not yet available for parenteral use.

- Halfan® (SmithKline & Beecham). Tablets 250 mg. Oral suspension 20 mg/ml.

References

1. Bryson HM, Goa KL (1992). Halofantrine. A review of its antimalarial activity, pharmacokinetic properties and therapeutic potential. *Drugs*, **43**, 236–258.
2. Karle JM, Olmeda R, Gerena L, Milhous WK (1994). *Plasmodium falciparum*: Role of absolute stereochemistry in the antimalarial activity of synthetic aminoalcohol antimalarial agents. *Exp Parasitol*, **76**, 345–351.
3. Keeratithakul D, Teja-Isavadharm P, Shanks GD *et al.* (1991). An improved high-performance liquid chromatographic method for the simultaneous measurement of halofantrine and desbutylhalofantrine in human serum. *Ther Drug Monit*, **13**, 64–68.
4. Mberu WM, Muhia DK, Watkins DK (1992). Measurement of halofantrine and its major meta-

bolite N-desbutylhalofantrine in plasma and blood by high-performance liquid chromatography: a new methodology. *J Chromatogr*, **581**, 156–160.
5. Milton KA, Edwards G, Ward SA, Orme M L'E, Breckenridge AM (1989). Pharmacokinetics of halofantrine in man: effects of food and dose size. *Br J Clin Pharmacol*, **28**, 71–77.
6. Karbwang J, Na Bangchang K (1994). Clinical pharmacokinetics of halofantrine. *Clin Pharmacokinet*, **27**(2), 104–119.
7. Krishna S, ter Kuile F, Supanaranond W, Pukrittayakamee S, Teja-Isavadharm P, Kyle D, White NJ (1993). Pharmacokinetic efficacy and toxicity of parenteral halofantrine in uncomplicated malaria. *Br J Clin Pharmacol*, **36**, 565–591.
8. Karbwang J, Milton KA, Na Bangchang K, Ward SA, Edwards G, Bunnag D (1991). Pharmacokinetics of halofantrine in Thai patients with acute uncomplicated *falciparum* malaria. *Br J Clin Pharmacol*, **31**, 484–487.
9. Karbwang J, Ward SA, Milton KA, Na Bangchang K, Edwards G (1991). Pharmacokinetics of halofantrine in healthy Thai volunteers. *Br J Clin Pharmacol*, **32**, 639–640.
10. Clerinx J, Taelman H (1993). Halofantrine treatment of uncomplicated *falciparum* malaria with high parasitaemia. *Trans R Soc Trop Med Hyg*, **87**, 80.
11. Watkins WM, Lury JD, Kariuki D, Koech DK, Oloo JA, Mosoba M, Mjomba M, Gilles HM (1988). Efficacy of multiple-dose halofantrine in treatment of chloroquine-resistant *falciparum* malaria in children in Kenya. *Lancet*, **2**, 247–250.
12. Salako LA, Sowunmi A, Walker O (1990). Evaluation of the Clinical trials and safety of halofantrine in *falciparum* malaria in Ibadan Nigeria. *Trans R Soc Trop Med Hyg*, **84**, 644–647.
13. Boudreau EF, Pang LW, Dixon KE, Webster HK, Pavanand K, Tosingha L, Somutsakorn P, Canfield CJ (1988). Malaria: Treatment efficacy of halofantrine (WR171,669) in initial field trials in Thailand. *Bull World Health Organ*, **66**, 227–235.
14. Richard-Lenoble D, Kombila M, Martz M, Gendrel D, Gendrel C, Moreno JL, Engohan E, Blanc G, Dupasquier I, Iannascoli F (1992). Efficacy safety and acceptability of halofantrine in the treatment of acute *Plasmodium falciparum* malaria in African children (Gabon). *J Trop Ped*, **38**, 7–11.
15. Wirima J, Molyneux ME, Khoromana C, Gilles HM (1988). Clinical trials with halofantrine hydrochloride in Malawi. *Lancet*, **340**, 250–251.
16. Coulaud JP, Le Bras J, Mathéron S, Morinière B, Saimot AG, Rossignol JF (1986). Treatment of imported cases of *falciparum* malaria in France with halofantrine. *Trans R Soc Trop Med Hyg*, **80**, 615–616.
17. Weinke T, Löscher T, Fleischer K, Kretschmer H, Pohle HD, Kohler B, Schlunk T, Clemens R, Bock HL (1992). The efficacy of halofantrine in the treatment of acute malaria in nonimmune travellers. *Am J Trop Med Hyg*, **47**, 1–5.
18. ter Kuile FO, Dolan G, Nosten F, Edstein MD, Luxemburger C, Phaipun L, Chongsuphajaisiddhi T, Webster HK, White NJ (1993). Halofantrine versus mefloquine in treatment of multidrug-resistant *falciparum* malaria. *Lancet*, **341**, 1044–1049.
19. Nosten F, ter Kuile FO, Luxemburger C, Woodrow C, Kyle DPE, Chongsuphajaisiddhi T, White NJ (1993). Cardiac effects of antimalarial treatment with halofantrine. *Lancet*, **341**, 1054–1056.
20. Ketrangsee S, Vijaykadga S, Yamokgul P, Jatapadma S, Thimasarn K, Rooney W (1992). Comparative trial on the response of Plasmodium falciparum to halofantrine and mefloquine in Trat Province Eastern Thailand. *Southeast Asian J Trop Med Public Health*, **23**, 55–58.

Ivermectin

Chemical structure

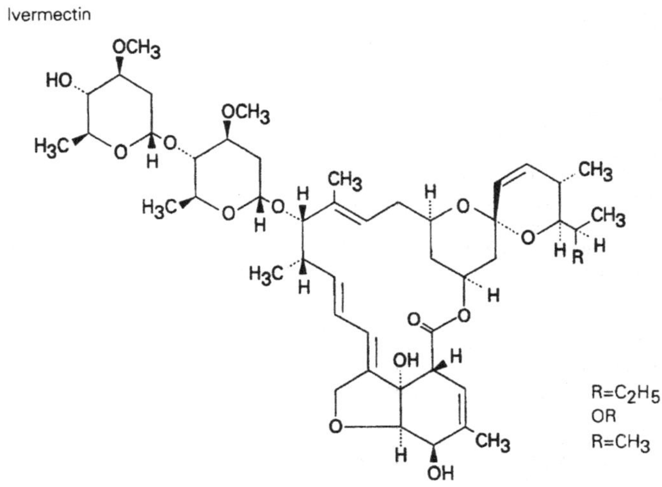

Ivermectin component B_{1a}:R = C_2H_5.
Ivermectin component B_{1b}:R = CH_3.
Ivermectin is a mixture of two closely related compounds with more than 80% of component B_{1a} and less than 20% of component B_{1b}. Neutral compound.

Physical properties

B_{1a}: MW 875; B_{1b}: MW 861. Practically insoluble in water.

Pharmacology and mechanism of action

Ivermectin belongs to a class of substances known as the avermectines. These are macrocylic lactones produced by fermentation of an actinomycete, *Streptomyces avermitilis*. Ivermectin is a broad spectrum agent active against nematodes and arthropods in domestic animals and is thus widely used in veterinary medicine (1). The drug was first introduced in man in 1981. It has been shown to be effective against a wide range of nematodes such as *Strongyloides sp.*, *Trichuris trichiura*, *Enterobius vermicularis*, *Ascaris lumbricoides*, hook worms and *Wuchereria bancrofti*. However, it has no effect against liver flukes and cestodes (2).

Presently it is regarded as the drug of choice in onchocerciasis. It is a potent microfilaricide, but it does not possess any significant macrofilaricidal effect (3). Between 2 to 3 days after oral administration, microfilariae in the skin start to disappear rapidly, while those in the cornea and the anterior chamber of the eyes are eliminated more gradually. This is an effect which lasts for up to 12 months (4–7). One month after administration, the microfilariae in the uterus of the worms are also affected where they get trapped and eventually degenerate and get resorbed (7). This long-term suppression of microfilariae has potential usefulness in interrupting the transmission of the disease (8, 9).

The mechanism of action of ivermectin against onchocerciasis is not clearly understood, but it is presumed to be a GABA-agonist. In susceptible organisms the drug acts by potentiating the release of gamma-aminobutyric acid (GABA) at postsynaptic sites on the neuromuscular junction rendering the nematode paralysed (10).

Pharmacokinetics

Specific HPLC methods have been described for the determination of ivermectin (11–13).

The absolute oral bioavailability of the drug is not known. Following oral administration of a single 12 mg dose of ivermectin to 12 healthy males, no difference in mean peak plasma concentrations between tablet and capsule formulations was observed (46 *vs* 50 μg/l). However, when ivermectin was administered in an aqueous ethanol solution the Cmax was virtually doubled and the relative bioavailability of the tablet was calculated to be 60% of that of the solution. Times to peak plasma levels were around 4 hours and did not differ between the formulations (14). Ivermectin has an apparent volume of distribution of around 48 l. Its elimination half-life after an oral solution was around 28 hours (14). About 93% of the drug is bound to plasma proteins (15). The metabolism of the drug in humans is not well investigated, however, it is reported to be hydroxylated and demethylated *in vitro* and *in vivo* in various animal species (16, 17).

Ivermectin is excreted through the bile and eliminated with the faeces. Less than 1% of the parent drug may be excreted with the urine (18).

Clinical trials

Ivermectin is one of the few drugs used against tropical parasitic infections that has been properly evaluated. The results of a large number of clinical trials ranging from phase I to IV have been summarized in two recent reviews (2, 3).

Early clinical trials have shown ivermectin to be an effective slow microfilaricide in patients with light infections of onchocerciasis in single oral doses between 30–50 μg/kg (6). Subsequently, open dose-finding studies in patients with moderate and heavy infections, including ocular involvement, have confirmed the initial findings after single oral doses of 50, 100, 150 or 200 μg/kg (6, 7, 19). In all these studies ivermectin eliminated microfilariae in the skin and eyes slowly, maintaining low levels of the parasite in those organs for about a year. In contrast to diethylcarbamazine (DEC), side effects were mild and transient and there were no severe ophthalmological adverse effects.

In phase II clinical studies, a number of randomized, double-blind studies with placebo controls and reference drug DEC were carried out in some West African countries including Senegal (20), Ghana (21), Mali (22) and Liberia (23). Only male patients were included in the studies. They had moderate to heavy skin microfilarial densities and almost all had mild or moderate eye lesions. Treatment consisted of ivermectin (as a single oral dose of 150 μg/kg) or DEC (approx. 0.8 mg/kg for 2 days, followed by 1.6 mg/kg twice daily for 6 days), or matching placebo capsules. The results were largely similar to those reported in the open studies. Both drugs gave prompt reductions in skin microfilariae counts to 0–4% of original counts by day 8 for DEC, and day 14 for ivermectin . However, the effect was more long lasting for the group which received ivermectin. One year after treatment microfilariae had risen to about 45% of pre-treatment counts (DEC group) or 9% of pre-treatment values (ivermectin group) (23). The effect of the drugs on the microfilariae in the eyes was more rapid after DEC treatment than with ivermectin. DEC eliminated microfilariae from the anterior chamber in 8 days, compared to 6 months after ivermectin treatment (21, 23). Systemic and ocular adverse reactions were fewer and less severe after ivermectin treatment.

Phase III studies with large numbers of patients have been conducted in Liberia, Ghana, Mali, Ivory Coast and Togo (2). They included both men and non-pregnant women. The dosages used were similar as above and the results were largely confirmatory of earlier reports.

In phase IV studies the drug was evaluated in the field under large scale community trials. Besides the confirmatory results of safety and efficacy, the major outcome from these studies was that ivermectin has a potential usefulness in interrupting the transmission of the disease from man to the vector. A review of these community based-studies has been published (24).

In lymphatic filariasis (*Wuchereria bancrofti* and *Brugia malayi*), experience is still limited. However, several studies comparing DEC (3 mg/kg for one day followed by 6 mg/kg for 12 days) versus ivermectin (single doses between 20 to 200 mg/kg) have reported DEC to be more rapidly acting in suppressing the microfilariae, but only for up to 3 to 6 months at which time most of those who were treated with ivermectin had microfilariae of 30–50% of the original levels (25–29). In one study (27), a single dose of 200 μg/kg per day for two days has been reported to have suppressed the microfilariae for up to one year. There is no evidence showing that ivermectin has any major effect on the adult filariae. Unlike onchocerciasis, the pathology of the infection in filariasis is largely due to the adult worm.

Indications

Ivermectin is the drug of choice against onchocerciasis. It is, however, an expensive drug and its distribution is still restricted. The role of ivermectin in lymphatic filariasis is not yet well investigated.

Pregnancy and lactation

Teratogenicity have not been reported in rats, but doses above 1.5 mg/kg were neurotoxic to the new borns during lactation (30, 31). Documentation in man is limited. In a large study in Liberia, where 14,000 patients were treated annually for 3 years, 203 children were born to women who received the drug during pregnancy. The occurrence of birth defects in children from treated mothers did not differ significantly from an untreated reference population (32). Because of the high risk of blindness due to onchocerciasis, and the lack of reports of teratogenicity despite the widespread use of the drug, ivermectin can be given after the first trimester.

The drug is excreted into breast milk and may attain concentrations of around 30% of that in plasma (33). Less than 10% of what goes into breast milk has been estimated to be taken up by the infant which has been regarded as clinically insignificant.

Side effects

About 1.5 million people, mainly in West Africa, have now been treated with ivermectin, and all the evidence indicates that it is a safe drug, which is suitable for large scale treatment programmes (32, 34, 35). Side effects reported include fever, itching, dizziness, oedema, mild Mazzotti reaction, and minimal ocular inflammation in patients with eye involvement. The side effects usually occur during the first 3 days after treatment and are dose dependent. The reported incidences of side effects vary. In one review (35), which covered 50,929 patients who had been treated with ivermectin, around 9% were reported to have suffered from side effects. The most frequent reaction was symptomatic postural hypotension. The authors reported that the incidence of side effects was directly proportional to the microfilarial density in the skin (35). In hyperendemic areas, a much higher incidence of adverse reactions may be seen (36).

Homeida *et al.* (37), reported a high incidence of prolongation of the prothrombin time in Sudanese patients treated with ivermectin, but this has not been confirmed in other studies (38–40).

Contraindications and precautions

Experience of the drug is lacking for children under 5 years of age. Since the drug acts by potentiating GABA, there is a concern that CNS effects may be seen in humans whose blood–brain barrier is impaired (e.g. by meningitis, trypanosomiasis). Ivermectin does not cross the blood–brain barrier; however, severe CNS toxicity has been reported from animals without a blood–brain barrier (e.g. collie dog) (41). The relevance of this in humans is not known.

Interactions

There are no reports of harmful drug interactions, but theoretically, the drug may potentiate the effects of other drugs that are agonists of the GABA receptors (e.g. benzodiazepines and sodium valproate).

Dosage (2)

Adults and children under 5 years
150 μg/kg as a single dose. Higher doses increase adverse reactions without an increase of efficacy. Annual re-treatment with this dosage is necessary, to ensure suppression of *O. volvulus* microfilariae. Patients with heavy ocular infection may require more frequent re-treatment, i.e. every 6 months.

Preparations

- Mectizan® (Merck Sharp & Dohme). Tablets 6 mg.

References

1. Campbell WC, Fisher MH, Stapley EO, Albers-Schönberg G, Jacob TA (1983). Ivermectin: a potent new antiparasitic agent. *Science*, **212**, 823–828.
2. Campbell WC (1991). Ivermectin as an antiparasitic agent for use in humans. *Annu Rev Microbiol*, **45**, 445–474.
3. Goa KL, McTavish D, Clissold SP (1991). Ivermectin: A review of its antifilarial activity, pharmacokinetic properties and clinical trials in onchocerciasis. *Drugs*, **42**, 640–658.
4. Awadzi K, Dadzie KY, Schultz-Key H, Haddock DRW, Gilles HM, Aziz MA (1984). Ivermectin in onchocerciasis. *Lancet*, **2**, 291.
5. Aziz MA, Diallo S, Diop IM, Larivière M, Porta M, Gaxotte P, Deluol AM, Cenac J (1982). Ivermectin in onchocerciasis. *Lancet*, **2**, 1456–1457.

6. Coulaud JP, Larivière M, Aziz MA, Gervais MC, Gaxotte P, Deluol AM, Cenac J (1984). Ivermectin in onchocerciasis. *Lancet*, **2**, 526–527.
7. Schultz-Key H (1990). Observations on the reproductive biology of *Onchocerca volvulus*. *Acta Leiden*, **59**, 27–43.
8. Taylor HR, Pacqué M, Munoz B, Greene BM (1990). Impact of mass treatment of onchocerciasis with ivermectin on the transmission of infection. *Science*, **250**, 116–118.
9. Cupp EW, Bernardo MJ, Kiszewski AE, Collins RC, Taylor HR, Aziz MA, Greene BM (1986). The effects of ivermectin on transmission of *Onchocerca volvulus*. *Science*, **231**, 740–742.
10. Pong S-S, Wang CC, Fritz LC (1980). Studies on the mechanism of action of avermectin B_{1a}: stimulation of release of gamma-aminobutyric acid from brain synaptosomes. *J Neurochem*, **34**, 351–358.
11. Chiou R, Stubbs RJ, Bayne WF (1987). Determination of ivermectin in human plasma and milk by high performance liquid chromatography with fluorescence detection. *J Chromatogr*, **416**, 196–202.
12. Dickinson CM (1990). Improved high-performance liquid chromatographic method for quantification of ivermectin in whole blood, serum or muscle tissue. *J Chromatogr*, **528**, 250–257.
13. Krishna DR, Klotz U (1993). Determination of ivermectin in human plasma by high-performance liquid chromatography. *Arzneimittelforschung*, **43**, 609–611.
14. Edwards G, Dingsdale A, Helsby N, Orme ML'E, Breckenridge AM (1988). The relative systemic availability of ivermectin after administration as capsule, tablet, and oral solution. *Eur J Clin Pharmacol*, **35**, 681–684.
15. Okonkwo PO, Ogbuokiri JE, Ofoegbu E, Klotz U (1993). Protein binding and ivermectin estimations in patients with onchocerciasis. *Clin Pharmacol Ther*, **53**, 426–430.
16. Chiu S-HL, Carlin JR, Sestokas E, Taub R, Buhs RP, Green M, *et al.* (1986). Metabolic disposition of ivermectin in tissues of cattle, sheep and rats. *Drug Metab Disp*, **14**, 590–600.
17. Chiu S-HL, Sestokas E, Taub R, Smith JL, Arison B, *et al.* (1984). The metabolism of avermectin-H_2B_{1a} by pig liver microsomes. *Drug Metab Disp*, **12**, 464–469.
18. Fink DW, Porras AG (1989). Pharmacokinetics of ivermectin in animals and man. In: Campbell WC, ed. *Ivermectin and abamectin*, (New York: Springer-Verlag), pp. 113–130.
19. Awadzi K, Dadzie KY, Schultz-Key H, Haddock DRW, Gilles HM, Aziz MA (1985). The chemotherapy of onchocerciasis X. An assessment of four single dose treatment regimes of MK–933 (ivermectin) in human onchocerciasis. *Ann Trop Med Parasitol*, **79**, 63–78.
20. Diallo S, Aziz MA, Larivière M, Diallo JS, Diop-Mar I, N'Dir O, Badiane S, Py D, Schulz-Key H, Gaxotte P, *et al.* (1986). A double blind comparison of the efficacy and safety of ivermectin in a placebo controlled study of Senegalese patients with onchocerciasis. *Trans R Soc Trop Med Hyg*, **80**, 927–934.
21. Dadzie KY, Bird AC, Awadzi K, Schultz-Key H, Gilles HM, Aziz MA (1987). Ocular findings in a double-blind study of ivermectin versus diethylcarbamazine versus placebo in the treatment of onchocerciasis. *Br J Ophtalmol*, **71**, 78–85.
22. Albiez EJ, Newland HS, White AT, Kaiser A, Greene BM, Taylor HR, Büttner DW (1988). Chemotherapy of onchocerciasis with high doses of diethylcarbamazine or a single dose of ivermectin: Microfilaria levels and side effects. *Trop Med Parasitol*, **39**, 19–24.
23. Larivière M, Vingtain P, Aziz M, Beauvais B, Weimann D, Derouin F, Ginoux J, Schulz-Key H, Gaxotte P, Basset D (1985). Double-blind study of ivermectin and diethylcarbamazine in African onchocerciasis patients with ocular involvement. *Lancet*, **2**, 174–177.
24. Remme J, De Sole G, Dadzie KY, Alley ES, Baker RH, Habbema JD, Plaisier AP, van Oortmarssen GJ, Samba EM (1990). Large scale ivermectin distribution and its epidemiological consequences. *Acta Leiden*, **59**, 177–191.
25. Diallo S, Aziz MA, Nadir O, Badiane S, Bah IB, Gaye O (1987). Dose-finding study of ivermectin in treatment of filariasis due to *Wuchereria bancrofti*. *Lancet*, **1**, 1030.
26. Kumaraswami V, Ottesen E, Vijayasekaran V, Uma Devi S, Swaminathan M, Aziz MA, Sarma GR, Prabhakar R, Tripathy S (1988). Ivermectin for the treatment of *Wuchereria bancrofti* filariasis. Efficacy and adverse reactions. *J Am Med Ass*, **259**, 3150–3153.
27. Richards Jr FO, Eberhard ML, Bryan RT, McNeely DF, Lammie PJ, McNeely MB, Bernard Y, Hightower AW, Spencer HC (1991). Comparison of high dose ivermectin and diethylcarbamazine for activity against bancroftian filariasis in Haiti. *Am J Trop Med Hyg*, **44**, 3–10.
28. Ottesen E, Vijayasekaran V, Kumaraswami V, Perumal Pillai SV, Sadanandam A, Phil M,

Fredrick S, Prabhakar R, Tripathy SP (1990). A controlled trial of ivermectin and diethylcarbamazine in lymphatic filariasis. *N Eng J Med*, **322**, 1113–1117.
29. Cartel J-L, Spiegel A, Nguyen L, Genelle B, Roux J-F (1991). Double blind study on efficacy and safety of single doses of ivermectin and diethylcarbamazine for treatment of Polynesian *Wuchereria bancrofti* carriers. Results at six months. *Trop Med Parasitol*, **42**, 38–40.
30. Poul J-M (1988). Effects of perinatal ivermectin exposure on behavioral development of rats. *Neurotoxicol Teratol*, **10**, 267–272.
31. Lankas GR, Minsker DH, Robertson RT (1989). Effects of ivermectin on reproduction and neonatal toxicity in rats. *Food Chem Toxicol*, **27**, 523–529.
32. Pacque M, Munoz B, Poetschke G, Foose J, Greene B, Taylor H (1990). Pregnancy outcome after inadvertent ivermectin treatment during community-based distribution. *Lancet*, **336**, 1486–1489.
33. Ogbuokiri JE, Ozumba BC, Okonkwo PO (1993). Ivermectin levels in human breast milk. *Eur J Clin Pharmacol*, **45**, 389–390.
34. Whitworth JAG (1992). Drug of the month: ivermectin. *Tropical doctor*, **22**, 163–164.
35. De Sole G, Remme J, Awadzi K, Accorsi S, Alley ES, Ba O, Dadzie KY, Giese J, Karam M, Keita FM (1989). Adverse reactions after large-scale treatment of onchocerciasis with ivermectin: combined study from eight community trials. *Bull WHO*, **67**, 707–719.
36. Whitworth JAG, Morgan D, Maude GH, Taylor DW (1988). Community-based treatment with ivermectin. *Lancet*, **2** 97–98.
37. Homeida MM, Bagi IA, Ghalib HW, el Sheikh HE, Ismail A, Yousif MA, Sulieman S, Ali HM, Bennett JL, Williams J (1988). Prolongation of prothrombin time with ivermectin. *Lancet*, **i**, 1346–1347.
38. Pacque MC, Munoz B, White AT, Williams PN, Greene BM, Taylor HR (1989). Ivermectin and prothrombin time. *Lancet*, **1**, 1140.
39. Whitworth JAG, Hay CRM, McNicholas AM, Morgan D, Maude GH, Taylor DW (1992). Coagulation abnormalities and ivermectin. *Ann Trop Med Parasitol*, **86**, 301–395.
40. Richards Jr FO, McNeely MB, Bryan RT, Eberhard ML, McNeely DF, Lamie PJ, Spencer HC (1989). Ivermectin and prothrombin time. *Lancet*, **i**, 1139–1140.
41. Campbell WC, Benz GW (1984). Ivermectin: a review of efficacy and safety. *J Vet Pharmacol Ther*, **7**, 1–16.

Levamisole

Chemical structure

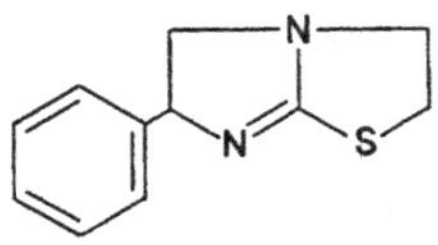

Physical properties

Base: MW 204; hydrochloride: MW 241; pKa 8.0. 1 g dissolves in 2 ml of water. Protect from light.

Pharmacology and mechanism of action

Levamisole is the L-isomer of tetramisole and is more active than the racemic mixture. It was introduced in 1966 as a veterinary drug and a little later as a human anthelminthic drug against ascariasis. The drug has also shown to be effective against hookworms (*Ancylostoma duodenale* and *Necator americanus*), but results of reported studies are inconsistent (1).

The mechanism of action of levamisole in helminthiasis is through its stimulation of autonomic ganglia (nicotinic receptors) of the worms. On exposure to the drug, immature and adult worms show spastic contraction followed by tonic paralysis. This mechanism seems to be common to other anthelminthics such as pyrantel and bephenium hydroxynaphthoate (2).

In higher doses, levamisole acts as an immunostimulant. It restores depressed cell-mediated immune mechanisms in peripheral T-lymphocytes, but may have marginal effects in immunologically competent individuals (3). The clinical implication of this effect in the treatment of helminthiasis is unknown.

Pharmacokinetics

A specific GC method have been described for the determination of levamisole (4).

The absolute bioavailability of the drug is unknown. After a single oral dose of 150 mg or 2.5 mg/kg in healthy volunteers, peak plasma levels of 0.5–0.7 μg/ml were reached within 2 hours. The apparent volume of distribution varied from 86 to 266 l (5). The drug is rapidly and extensively metabolised. One metabolite, hydroxylevamisole, has been identified in the urine of man and rats, but several other unidentified metabolites are thought to be formed (6). In rats, another metabolite, OMPI (2-oxo-3-(2-mercaptoethyl)-5-phenylimidazolidine) with immunotropic properties has been identified (7).

Excretion is rapid with a plasma elimination half-life between 4 and 5 hours (5, 6).

Following the administration of tritium labelled levamisole to 3 healthy volunteers, approximately 60% of the dose was excreted with the urine within the first 24 hours as hydroxylevamisole, largely as conjugates with glucuronic acid. Between 3% and 6% of the dose was excreted as the parent drug. Only 4% of the radioactivity was recovered in the faeces (6). The excretion of the drug is inversely proportional to urinary pH (6).

Clinical trials

In an open dose finding study in Indonesia, 333 patients with ascariasis were treated with levamisole 2.5–5 mg/kg for 1–3 days. After 7–10 days, the cure rate was above 87% (8). In another open study conducted in Nigeria, 199 children with ascariasis were treated with levamisole 4–10 mg/kg to a maximum dose of 240 mg. Treatment regimens included a single dose given for 1 to 2 days and weekly doses given for 2 to 3 consecutive weeks. After 7–14 days, high cure rates (above 95%) were recorded from all patients and the single dosage regimen proved to be as effective as the repeated dosage regimen (9). In a multicentre clinical study carried out in Brazil, Iran, and United States, 914 patients with ascariasis were treated with single doses of levamisole (2.5–5 mg/kg, maximum 150 mg) or piperazine (150 mg/kg, maximum 3.5 g). 3–6 weeks post treatment, the cure rates were 92% for levamisole, and 60% for piperazine (10). Similar results have been reported by other studies (11, 12).

Against ancylostomiasis, single dose regimens of 2.5 mg/kg or 150 mg of levamisole cured 64 to 93% of the patients (13, 14). In another study of 50 patients with ancylostomiasis, 41 received a single dose of 5 mg/kg while the remaining 9 were given two doses of 5 mg/kg of levamisole, the second dose being given 2 days after the first. A 100% cure rate was reported in both groups 7 days after treatment. No serious side effects were reported to be associated with this dosage regimen (15). Against *Necator americanus*, cure rates reported from various studies using similar dose regimens were inconsistent and varied between 0 and 84% (8, 9).

Indications

Monoinfections with *Ascaris lumbricoides*. In polyinfections, mebendazole is the drug of choice.

Pregnancy and lactation

Teratogenicity has not been reported in rats and rabbits treated with doses between 5 and 150 mg/kg of levamisole during pregnancy (16). Documentation in man is lacking. Treatment with levamisole should be postponed until delivery, unless there is a strong indication for use.

Its excretion into breast milk is unknown.

Side effects

During the treatment of nematode infections the drug produces minor side effects including nausea, vomiting, abdominal pain and headache (12, 13). During prolonged treatment as an immunomodulator in rheumatic arthritis and in cancer patients, serious side effects such as blood disorders (agranulocytosis, neutropenia and thrombocytopenia), kidney damage, influenza-like reactions, vasculitis, photosensitivity and allergy to the drug have been reported (7, 16).

Contraindications and precautions

The drug should be avoided in patients allergic to the drug. Administration of levamisole may provoke a reaction similar to that seen after intake of alcohol together with disulfiram. During long-term treatment, patients with kidney damage or with blood disorders may experience exacerbation of their diseases.

Interactions

Levamisole has been reported to displace the protein binding of rifampicin *in vitro* (17). The clinical significance of this is as yet unknown.

Dosage

Ascariasis

Adults
150 mg levamisole (base) as a single dose.
Children
2.5 mg/kg levamisole (base) as a single dose.

Preparations

Available as levamisole hydrochloride: 118 mg is equivalent to 100 mg base.

- Ketrax® (Zeneca). Oral solution 40 mg base per 5 ml. Tablets 40 mg base.
- Solaskil® (Rhône-Poulenc Rorer). Tablets 30 mg base, 150 mg base.
- Ergamisol® (Lederle). Tablets 50 mg base.
- Levamisol® (Janssen). Tablets 50 mg base.

References

1. Miller MJ (1980). Use of levamisole in parasitic infections. *Drugs*, **19**, 122–130.
2. van Wauwe J, Janssen PAJ (1991). On the biochemical mode of action of levamisole: an update. *Int J Immunopharmacol*, **13**, 3–9.
3. Renoux G (1980). The general immunopharmacology of levamisole. *Drugs*, **19**, 89–99.
4. Kouassi E, Caillé G, Léry L, Larivière L, Vézina M (1986). Novel assay and pharmacokinetics of levamisole and p-hydroxylevamisole in human plasma and urine. *Biopharmaceut Drug Dispos*, **7**, 71–89.
5. Lucykx M, Rousseau F, Cazin M, Brunet C, Cazin JC, Haguenoer JM, Devulder B, Lesieur I, Lesieur D, Gosselin P, Adenis L, Cappelaere P, Demaille A (1982). Pharmacokinetics of levamisole in healthy subjects and cancer patients. *Eur J Drug Metab Pharmacokinet*, **7**, 247–254.
6. Adams JG (1978). Pharmacokinetics of levamisole. *J Rheumatol*, **5**, 137–142.
7. Chrisp P, McTavish D (1991). Levamisole/fluorouracil: A review of their pharmacology and adjuvant therapeutic use in colorectal cancer. *Drugs & Aging*, **14**, 317–337.
8. Thienpoint D, Brugmans J, Abadi K, Tanamal S (1969). Tetramisole in the treatment of nematode infections. *Am J Trop Med Hyg*, **18**, 520–525.
9. Lucas AO, Oduntan SO (1972). Treatment of hookworm infection and other parasites with l-tetramisole (Ketrax). *Ann Trop Med Parasitol*, **66**, 391–398.
10. Miller MJ, Farahmandian I, Arfaa F, Katz N, Winsor E, Bennett E (1978). An evaluation of levamisole for treatment of ascariasis. *South Med J*, **71**, 137–140.
11. Moens M, Dom J, Burke WE, Schlossberg S, Schuermans V (1978). Levamisole in ascariasis. A multicentre controlled evaluation. *Am J Trop Med Hyg*, **27**, 897–904.
12. Lionel ND, Mirando EH, Nanayakkara JC, Soysa PE (1969). Levamisole in the treatment of ascariasis in children. *BMJ*, **4**, 340–341.

13. Farid Z, Bassily S, Miner WF, Hassan A, Laughli LW (1977). Comparative single-dose treatment of hookworm and roundworm infections with levamisole, pyrantel and bephenium. *J Trop Med Hyg*, **80**, 107–108.
14. Huys J, van den Berghe G, Freyens P, Kayihigi J (1976). Treatment of ancylostomiasis with levamisole. *Afr J Med Sci*, **5**, 75–77.
15. Al-Saffar G, Al-Saleem M, Bakhous IJ (1971). L-tertramisole in the treatment of ancylostomiasis. *Trans R Soc Trop Med Hyg*, **65**, 836–837.
16. Amery WK, Butterworth BS (1983) The dosage regimen of levamisole in cancer: is it related to efficacy and safety? *Int J Immunopharmacol*, **5**, 1–9.
17. Pérez-Gallardo L, Blanco ML, Soria H, Escanero JF (1992). Displacement of rifampicin bound to serum proteins by addition of levamisole. *Biomed Pharmacother*, **46**, 173–174.

Mebendazole

Chemical structure

Mebendazole

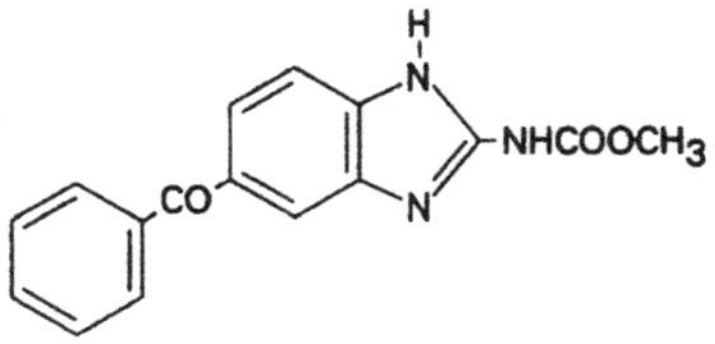

Physical properties

MW 295; pKa not known. Practically insoluble in water.

Pharmacology and mechanism of action

Mebendazole is a benzimidazole derivative with a broad spectrum of anthelminthic activity. It is highly effective against adult and larval stages of *Ascaris lumbricoides*, *Enterobius vermicularis*, *Trichuris trichiura*, hookworms (*Ancylostoma duodenale* and *Necator americanus*) and *Capillaria philippinensis*. It is also ovicidal against *Ascaris lumbricoides* and *Trichuris trichuria* (1). With high doses, the drug has some effect against hydatid disease (2). Recent *in vitro* studies have reported mebendazole to be more effective than metronidazole in killing *Giardia lamblia* (3, 4); however, clinical findings are inconclusive (5, 6, 7).

The mechanisms of action of benzimidazoles are similar. These drugs appear to bind to parasite tubules with subsequent inhibition of the polymerization of tubules to microtubules which is vital for the normal functioning of the parasite cells (8).

Pharmacokinetics

Specific HPLC methods using UV (9, 10) or electrochemical (11) detection have been described for mebendazole and its metabolites.

Mebendazole is taken orally. The oral bioavailability of the drug is less than 20% (12). However, its absorption can be increased several fold if taken with a fatty meal (13). Peak plasma levels are reached within 4 hours, but large intra- and inter-individual variability have been reported (13, 14). Its volume of distribution is around 1.2 l/kg (12). About 95% of the drug is bound to plasma proteins (14). It is extensively metabolized in the body largely to inactive metabolites (hydroxy- and aminometabolites), which have lower rates of clearance than the parent drug (15). The plasma elimination half-life of mebendazole is around 1 hour (12). The drug and its metabolites are excreted via the bile into the faeces (15). Small amounts are excreted with the urine.

Clinical trials

In open clinical studies on patients with ascariasis, cure rates of 95–100% were reported after treatment with mebendazole 100 mg twice daily for 3 days (16–19). Similar cure rates have been achieved in patients infected with *Enterobius vermicularis* (20, 21), *Necator americanus* (15) and *Ancylostoma duodenale* (16, 17) using the same dose regimen. In *Trichuris trichiura* infection, cure rates can vary from 45 to 100% (22, 23). In a double-blind placebo controlled study in Indonesia, a single dose of 500 mg mebendazole, has been reported to be effective, inexpensive and convenient when used in mass treatment programmes against soil transmitted nematodes (24).

High doses (40–70 mg/kg) over a long period administered to patients with hydatid disease have been shown to have good effects (25–28). However, most studies have reported albendazole to be more effective and safer than mebendazole, and it is today considered to be a better alternative (*see also* under Albendazole – Clinical trials).

Mebendazole has been studied in the therapy of onchocerciasis (*see also* under Diethylcarbamazine – Clinical trials). In large oral doses (2–3 g daily) mebendazole appears to sterilize adult worms, producing a gradual decline in skin microfilariae count; however, the effect is not permanent (29, 30).

Indications

Mebendazole is the drug of choice for mixed nematode infections due to *Trichuris trichiura*, *Ascaris lumbricoides*, *Enterobius vermicularis*, *Capillaria philippinensis* or hookworms.

The drug may be used against hydatid disease when albendazole is not available.

Pregnancy and lactation

Mebendazole in high doses is teratogenic and embryotoxic in rats (31). Documentation in man is lacking, despite the widespread use of the drug. Treatment with mebendazole should be avoided during early pregnancy.

Its excretion into breast milk is unknown.

Side effects

Despite the widespread use of the drug, few side effects have been reported, especially in patients with heavy infections. These include transitory abdominal pain, diarrhoea and slight headache. High doses of the drug such as those used in the treatment of hydatid disease have been associated with bone marrow toxicity, alopecia, hepatitis, glomerulonephritis, fever and exfoliative dermatitis (25–28).

Contraindications and precautions

When high doses of mebendazole are given, regular monitoring of serum-transaminase levels and leukocyte and platelet counts must be carried out. In patients with liver impairment dosage reductions must be made.

Interactions

The concomitant administration of phenytoin or carbamazepine has been reported to lower the plasma concentration of mebendazole (28), while cimetidine had the opposite effect (32).

Dosage

Infections with Ancylostoma duodenale, Ascaris lumbricoides, Necator americanus, Trichuris trichiura

Adults and children

100 mg twice daily for 3 days.

Infections with Enterobius vermicularis

Adults and children

A single dose of 100 mg repeated after 2 weeks.

Infections with Capillaria philippinensis

Adults and children

200 mg/kg twice daily for 21 days.

Hydatid disease

40 mg/kg daily for 1–6 months. Albendazole is the drug of choice and mebendazole should only be used if the former is not available.

Preparations

- Pantelmin® (Janssen). Oral solution 20 mg/ml. Tablets 100 mg, 500 mg.
- Vermox® (Janssen). Oral suspension 20 mg/ml. Tablets 100 mg, 500 mg.

Several other preparations are available.

References

1. Van den Bossche H, Rochette F, Horig C (1982). Mebendazole and related anthelminthics. *Adv Pharmacol Chemother*, **19**, 287–296.
2. Todorov T, Vutova K, Mechkov G, Georgiev P, Petkov D, Tonchev Z, Nedelkov G (1992). Chemotherapy of human cystic echinococcosis: comparative efficacy of mebendazole and albendazole. *Ann Trop Med Parasitol*, **86**, 59–66.
3. Cedillo-Rivera R, Munoz O (1992). *In-vitro* susceptibility of *Giardia lamblia* to albendazole, mebendazole and other chemotherapeutic agents. *J Med Microbiol*, **37**, 221–224.
4. Edlind TD, Hang TL, Chakraborty PR (1990). Activity of the anthelminthic benzimidazoles against *Giardia lamblia in vitro*. *J Infect Dis*, **162**, 1408–1411.
5. Al-Waili D, Al-Waili B, Saloom K (1988). Therapeutic use of mebendazole in giardial infections. *Trans R Soc Trop Med Hyg*, **82**, 438.
6. Al-Waili NSD, Hasan NU (1992). Mebendazole in giardial infections: A comparative study with metronidazole. *J Infect Dis*, **165**, 1170–1171.
7. Gascon J, Moreno A, Valls ME, Miro JM, Corachan M (1989). Failure of mebendazole treatment in *Giardia lamblia* infection. *Trans R Soc Trop Med Hyg*, **83**, 647.
8. Lacey E (1990). Mode of action of Benzimidazoles. *Parasitology Today*, **6**, 112–115.
9. Allan RJ, Goodman HT, Watson TR (1980). Two high-performance liquid chromatographic determinations for mebendazole and its metabolites in human plasma using a rapid Sep Pak C18 extraction. *J Chromatogr*, **183**, 311–319.
10. Ramanathan S, Nair NK, Mansor SM, Navaratnam V (1993). Determination of a new antifilarial drug, UMF–058, and mebendazole in whole blood by high-performance liquid chromatography. *J Chromatogr*, **615**, 303–307.
11. Betto P, Gianbenedetti M, Ponti F, Ferretti R, Settim G, Gargiulo M, Lorenzini R (1991). Application of a high-performance liquid chromatography coulometric method for the estimation of mebendazole and its metabolites in human sera. *J Chromatogr*, **563**, 115–123.

12. Dawson M, Braithwaite PA, Roberts MS, Watson TR (1985). The pharmacokinetics and bioavailability of a tracer dose of (3H)-mebendazole in man. *Br J Clin Pharmacol*, **19**, 79–86.
13. Münst GJ, Karlaganis G, Bircher J (1980). Plasma concentrations of mebendazole during treatment of echinococcosis. *Eur J Clin Pharmacol*, **17**, 375–378.
14. Braithwaite PA, Roberts MS, Allan RJ, Watson TR (1982). Clinical pharmacokinetics of high dose mebendazole in patients treated for cystic hydatid disease. *Eur J Clin Pharmacol*, **22**, 161–169.
15. Gottschall DW, Theodorides VJ, Wang R (1990). The metabolism of Benzimidazoles. *Parasitology Today*, **6**, 115–120.
16. Hutchison JG, Johnston NM, Plevey MV, Thangkhiew I, Aidney C (1975). Clinical trial of mebendazole, a broad-spectrum anthelminthic. *BMJ*, **2**, 309–310.
17. Chavarria AP, Swartzwelder JC, Villarejos VM, Zeledon R (1973). Mebendazole, an effective broad-spectrum anthelminthic. *Am J Trop Med Hyg*, **22**, 592–595.
18. Wolfe MS, Wershing JM (1974). Mebendazole: treatment of trichuriasis and ascariasis in Bahamian children. *J Am Med Ass*, **230**, 1408–1411.
19. Wagner ED, Rexinger DD (1978). *In vivo* effects of mebendazole and levamisole in the treatment of trichuriasis and ascariasis. *Am J Trop Med Hyg*, **27**, 203–205.
20. Brugmans JP, Thienpont DC, van Wijngaarden I, Vanparijs OF, Schuermans VL, Lauwers HL (1971). Mebendazole in enterobiasis: radiochemical and pilot clinical study in 1278 subjects. *J Am Med Ass*, **217**, 313–316.
21. Lormans JAG, Wesel AJT, Vanparus OF (1975). Mebendazole (R 17635) in enterobiasis. A clinical trial in mental retardees. *Chemotherapy*, **21**, 255–260.
22. Blechman MG (1975). Clinical effectiveness of mebendazole in the treatment of trichuriasis. *Curr Ther Res*, **18**, 800–803.
23. Scragg JN, Proctor EM (1977). Mebendazole in the treatment of severe symptomatic trichuriasis in children. *Am J Trop Med Hyg*, **24**, 932–984.
24. Abadi K (1985). Single dose mebendazole therapy for soil-transmitted nematodes. *Am J Trop Med Hyg*, **34**, 129–133.
25. Wilson JF, Rausch RL, McMahon BJ, Schantz PM (1992). Parasitological effect of chemotherapy in alveolar hydatid disease: Review of experience with mebendazole and albendazole in Alaskan eskimos. *Clin. Infect Dis*, **15**, 234–249.
26. Ellis M, von Sinner W, Al-hokail A, Siek JA (1992). Clinical-radiological evaluation of benzimidazoles in the management of *Echinococcus granulosus* cysts. *Scand J Infect Dis*, **24**, 1–13.
27. Todorov T, Vutova K, Mechkov G, Tonchev Z, Georgiev P, Lazarova I (1992). Experience in the chemotherapy of severe, inoperable echinococcosis in man. *Infection*, **20**, 23–24.
28. Luder PJ, Siffert B, Witassek F, Meister F, Bircher J (1986). Treatment of hydatid disease with high oral doses of mebendazole. Long-term follow-up of plasma mebendazole levels and drug interactions. *Eur J Clin Pharmacol*, **31**, 443–448.
29. Awadzi K, Schulz-Key H, Edwards G, Breckenridge A, Orme M'LE, Gilles H (1990). The chemotherapy of onchocerciasis XIV. Studies with mebendazole citrate. *Trop Med Parasitol*, **41**, 384–386.
30. Rivas-Alcala AR, Mackenzie GD, Gomez-Rojo E, Greene M, Taylor HR (1984). The effects of diethylcarbamazine, mebendazole, levamisole on *Onchocerca volvulus in vitro* and *in vivo*. *Tropenmed Parasitol*, **35**, 71–77.
31. Mebendazole. *Therapeutic Drugs*, edited by Sir Colin Dollery (1990), (London: Churchill Livingstone), pp. M12–M16.
32. Bekhti A, Pirotte J (1987). Cimetidine increases serum mebendazole concentrations. Implications for treatment of hepatic hydatid cysts. *Br J Clin Pharmacol*, **24**, 390–392.

Mefloquine

Chemical structure

Mefloquine

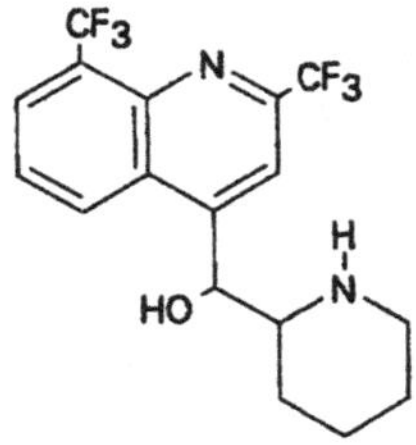

Physical properties

Base: MW 374; hydrochloride: MW 415; pKa: 8.6. The salt is slightly soluble in water.

Pharmacology and mechanism of action

Mefloquine is a quinolinemethanol derivative which is structurally related to quinine. It was synthesized and tested by the United States army in the 1960s. The available mefloquine preparation is a racemate with two enantiomers in equal proportions (1). It was introduced for the treatment of multiresistant *P. falciparum* in the mid-1980s (2). In Africa, there are only occasional reports of therapeutic failures, but there has been a rapid development of resistance in the 1990s in parts of Southeast Asia, particularly in Thailand (3). Cross-resistance with quinine and halofantrine has also been reported (4, 5). The mechanism of action is not well established. Mefloquine is a schizontocidal drug active against the erythrocytic stages of all species of malaria parasites. It is inactive against exoerythrocytic forms and thus cannot prevent relapse of *P.vivax* and *P. ovale* infections (6).

Pharmacokinetics

HPLC methods have been described for determination of mefloquine (7) and its main metabolite 2,8-bis (trifluoromethyl)-4-quinolinecarboxylic acid (8). One method can determine mefloquine by using capillary blood (100 μl) dried on filter paper (9).

Mefloquine is a local irritant, and no parenteral preparation is available. Therefore, oral bioavailability, volume of distribution and clearance cannot be accurately determined. The relative oral bioavailability of the present Lariam® preparation has been improved compared to previous preparations. Another oral preparation (Mephaquin®, Mepha Pharmaceuticals) is produced, but whether it has a similar bioavailability has not yet been fully established. The absorption of mefloquine has been reported to be linear after single doses up to 1500 mg (10, 11). In one study with frequent sampling, an absorption half-life of approximately 1 hour was found and peak concentrations in plasma were reached after 2–12 hours (11).

The apparent volume of distribution has been calculated to 16–23 l/kg (11–14). The mean terminal half-life in healthy European, African, Brazilian, and Thai volunteers is 15–27 days with an inter-individual variation from 6–33 days (11–17). In Chinese volunteers given mefloquine in combination with sulphadoxine and pyrimethamine, the mean half-life was only 11 days and it is still not clear whether there are significant ethnic differences in pharmacokinetic parameters (18). In a study with 250 mg mefloquine given once weekly, steady-state concentrations were reached after 6–8 weeks (15). The pharmacokinetics do not seem to be different in children (19, 20).

Mefloquine is extensively (98%) bound to plasma proteins and also to tissues and red cell membranes (21, 22). The concentrations in plasma and whole blood are rather similar (14). In patients with cerebral malaria, mefloquine has not been detectable in the cerebrospinal fluid probably due to its high protein binding in plasma (23).

In the rat, mefloquine is metabolized in the liver, and the main metabolite in blood and urine is the corresponding carboxylic acid (24). This metabolite has also been regarded as the main metabolite in man as its area under the concentration versus time curve (AUC) in plasma is 3–4 times higher than the AUC for the parent compound (11, 12, 14). The peak plasma concentration of the main metabolite is reached within 4–14 days and the half-life is grossly similar to that of mefloquine (20). The metabolite has no significant antimalarial effect *in vitro* (25). At steady-state, 9% of the dose is found unchanged in the urine and 4% as the main metabolite (26). Animal studies have suggested that mefloquine undergoes enterohepatic circulation. However, direct evidence in man is still lacking.

Clinical trials

The outcome of antimalarial therapy is dependent upon the parasite susceptibility and the immunity of the patient (non- or semi-immune). Strains resistant to mefloquine are known to have been present both in Thailand and East Africa before the drug was deployed (27, 28).

Initial dose finding studies have shown that ≤1000 mg given to non-immune volunteers with *P. falciparum* malaria initially cleared the parasites and symptoms, but the parasites later frequently re-appeared (RI response) (29). In 673 semi-immune patients with *P. falciparum* malaria from several clinical studies, there was no obvious difference in efficacy between adults given 500, 750 or 1000 mg or between children given 20, 25, or 30 mg/kg (30). In an early study from Zambia, in adult (semi-immune) patients with *P. falciparum* malaria, a dose of 1000 mg mefloquine was effective (31). In Malawi, in children under 5 years of age with *P. falciparum* malaria given either 25 mg or 15 mg mefloquine/kg, the clinical and parasitological response was similar in both groups (32). In West Africa, a 100% cure rate has been reported in children with *P. falciparum* malaria after a single mefloquine dose of 25 mg/kg (33, 34).

In contrast to the African experience where mefloquine is not yet widely used, a rapid decrease in sensitivity has been reported from different parts of Thailand (Trat and Tak provinces) since the late 1980s when this drug became the first line treatment for *falciparum* malaria (17, 35). In other areas of Southeast Asia except Cambodia and parts of Myanmar, the situation is less alarming.

There are only a few reports of *P. falciparum* resistance from Africa in persons with prophylactic mefloquine concentrations where determination of mefloquine concentration has excluded poor compliance as the reason for failure. Previously it was recommended to prolong the dose interval to once every second week after the third week to prevent toxic accumulation. In 1991, a study in US peace corps volunteers in West Africa reported a high number of *P. falciparum* infections with this regimen (36). One reason was probably poor compliance, but the recommendation was generally to change the dose to 250 mg once weekly.

For other plasmodia, in particular *P. vivax*, parasitemia may develop after completion of prophylaxis due to the latent liver stage (37, 38).

Mefloquine has been combined with sulphadoxine/pyrimethamine (Fansimef®) with the aim to delay the appearance of resistance. However, resistance to sulphadoxine/pyrimethamine is already of high grade in areas with reduced mefloquine susceptibility and this combination is no longer recommended (2).

Indications

Treatment and prophylaxis against chloroquine resistant *P. falciparum* malaria.

Pregnancy and lactation

Mefloquine is neither mutagenic, carcinogenic, nor teratogenic in laboratory animals (39). There have been no maternal or fetal toxicity reported in two studies on the safety and efficacy of mefloquine prophylaxis during the second and third trimester of pregnancy (40, 41). However, the potential for teratogenic effects in the first trimester has not yet been resolved, and the drug should be avoided for prophylaxis during this period. It can still be given for treatment if other alternatives are not available or are less effective.

Mefloquine is excreted into the breast milk, but the amount of drug expected to be taken up by the infant is low (42).

Side effects

The severity and frequency of side effects during treatment are dose-related. In early clinical trials (a total of 436 patients), the most frequent adverse reactions were nausea (18%), diarrhoea (15%), dizziness (15%), vomiting (13%), sinus bradycardia (9%), abdominal pain (8%), skin itching or rash (1%) and behaviour disorders with paranoid ideas and hallucinations (1%) (30). Early vomiting within 1 hour after drug administration reduced the mefloquine concentrations in patients with *P. falciparum* malaria. This indicates that vomiting within 1 hour requires a repeated dose (43).

The most important side effects of mefloquine are neuropsychiatric reactions. In 7 volunteers given 15 mg mefloquine/kg, all experienced some neurological symptoms (concentration difficulties, dizziness, vertigo) within 6 hours after administration of the drug (44). Serious neuropsychiatric adverse reactions, in particular general convulsions, confusion, and hallucinations, have been reported after therapeutic use of mefloquine (45, 46). The incidence may be as high as 1% and the onset usually occurs within 4 days of intake (47). Neurological and psychiatric reactions seem to be dose-dependent (2, 47). All neuropsychiatric reactions are reversible once the drug administration is discontinued. No fatalities have been reported (48). Less frequent adverse effects mainly associated with curative doses are anorexia, asthenia, irregular heart rate, pulse irregularities, constipation, insomnia, diarrhoea, arthralgia, and hearing disturbances (49). Single case reports of Stevens–Johnson syndrome, severe facial rash, and agranulocytosis have been filed (50).

During prophylaxis, the frequency of reported symptoms among 2780 travellers using mefloquine was similar to that of chloroquine and only one possible serious reaction (depression) was reported (51). In 1991, the recommended prophylactic dose of mefloquine for adults was increased to 250 mg once weekly during the whole period. The experience with this dose is still limited, but it was well tolerated in a trial in US peace corps volunteers in West Africa (52). There are no large, double-blind, prospective, randomized studies that compare the risk for adverse reactions between different antimalarial drugs used for prophylaxis. Large retrospective studies in US Peace Corps volunteers and European travellers found no major difference in the incidence of side effects in mefloquine compared to chloroquine users (51, 52). Until May 1991, a total of 59 serious neuropsychiatric adverse reactions (26 convulsions, 12 depressions, 20 psychotic episodes, and one toxic encephalopathy) have been reported to Roche after prophylaxis (48) The majority (80%) of all neuropsychiatric reactions appeared within 3 weeks of onset of prophylaxis (53). In Germany, the risk of moderate to severe neuropsychiatric reactions during prophylaxis was calculated to be one in 13,000 users (54).

Contraindications

There is little experience in children under 2 years, and mefloquine is therefore not recommended in this group unless other alternatives are ineffective (3). Mefloquine should not be used by persons involved in activities requiring fine co-ordination and spatial performance such as air crews or by persons with a history of epilepsy or psychiatric disorders (3). People taking cardioactive drugs like digoxin, quinidine, beta-blockers, or calcium channel blockers should avoid mefloquine (3). Those on mefloquine prophylaxis or given mefloquine treatment should only be given quinine under close medical supervision because of the risk of additive neurological and cardiological toxicity (3).

Drug interactions

Sinus bradycardia and sinus arrhythmia are often seen during mefloquine treatment, and pharmacodynamic interactions can be anticipated when cardioactive drugs like quinine, quinidine, beta-blockers, or calcium channel blockers are used concomitantly.

Dosage

In the US, doses are expressed in terms of the hydrochloride (250 mg hydrochloride is equivalent to 228 mg base) while in the rest of the world they are usually expressed in terms of the base – see *Preparations*.

Treatment

The recommended treatment dose varies between 15 and 25 mg base/kg with a maximal total dose of ≤1500 mg. Ideally, the dose should be adjusted according to the parasite susceptibility and the host immunity – a low dose (15 mg/kg) would then be given to the semi-immune indigenous population in tropical Africa and a high dose (25 mg/kg) to all patients with *P. falciparum* malaria contracted in Southeast Asia.

WHO (55)
Adults and children
15 mg base/kg to a maximum of 1000 mg in two divided doses (radical cure may not be achieved in non-immune subjects weighing more than 60 kg).

UK (50)
Adults and children
20 mg base/kg to a maximum of 1500 mg.

US (50)
Adults
1250 mg base as a single dose.

France (56)
Adults
1250–1500 mg base divided in two or three doses.
Children
25 mg base/kg as a single dose.

Prophylaxis

The intake should start 1 week before travel and continue for 4 weeks after leaving the area. Recommended dose is 250 mg base once weekly to adults (3). Children are given 5 mg base/kg once weekly. Previously it has been recommended not to give mefloquine for longer periods than 3 months due to lack of experience, but this restriction has now been abolished (3, 57).

Preparations

Available as mefloquine hydrochloride: 274 mg hydrochloride is equal to 250 mg base.

- Lariam® (Roche) (or Laricum in some countries). Tablets 250 mg base (228 mg base in the US).
- Mephaquin® (Mepha). Tablets 250 mg base.

References

1. Sweeney T (1981). The present status of malaria chemotherapy: mefloquine, a novel antimalarial. *Med Res Rev*, **1**, 281–301.
2. *Practical Chemotherapy of Malaria*. Technical report series No. 805 (1990). (Geneva: World Health Organization).
3. International Travel and Health (1994). (Geneva: World Health Organization).
4. Rojas-Rivero L, Gay F, Bustos MDG, Ciceron L, Pichet C, Danis M, Gentilini M (1992). Mefloquine-halofantrine cross-resistance in *Plasmodium falciparum* induced by intermittant mefloquine pressure. *Am J Trop Med Hyg*, **47**, 372–377.
5. Brasseur P, Kouamouo J, Druilhe P (1991). Mefloquine-resistant malaria induced by inappropriate quinine regimens. *J Infect Dis*, **164**, 625–626.
6. Black RH, Canfield CJ, Clyde DF, Peters W, Wernsdorfer WH (1986). Mefloquine. In: *Chemotherapy of Malaria*, 2nd edn, edited by L. Bruce-Chwatt. (Geneva: World Health Organization 1986).
7. Grindel JM, Tilton PF, Shaffer RD (1977). Quantification of the antimalarial agent mefloquine in blood, plasma and urine using high-pressure liquid chromatography. *J Pharm Sci*, **66**, 834–837.
8. Bergqvist Y, Churchill FC (1988). Detection and determination of antimalarial drugs and their metabolites in body fluids. *J Chromatogr*, **434**, 1–20.
9. Mount DL, Churchill FC, Bergqvist Y (1990). Determination of mefloquine in blood, filter paper-absorbed blood and urine by 9-fluorenylmethyl chloroformate derivatization followed by liquid chromatography with fluorescence detection. *J Chromatogr*, **564**, 181–193.
10. Desjardins RE, Pamplin CL III, Bredow J, von Barry KG, Canfield CJ (1979). Kinetics of a new antimalarial, mefloquine. *Clin Pharmacol Ther*, **26**, 372–379.
11. Schwartz DE, Eckert G, Hartmann D, Weber B, Richard-Lenoble D, Ekue JMK, Gentilini M

(1982). Single dose kinetics of mefloquine in man. Plasma levels of the unchanged drug and of one of its metabolites. *Chemotherapy*, **28**, 70–84.
12. De Souza JM, Sheth UK, De Oliveira, RMG, Roulet H, De Souza SD (1985). An open randomized, phase III clinical trial of mefloquine and of quinine plus sulphadoxine-pyrimethamine in the treatment of symptomatic *falciparum* malaria in Brazil. *Bull World Health Organ*, **63**, 603–609.
13. Karbwang J, Bunnag D, Breckenridge AM, Back DJ (1987). The pharmaco-kinetics of mefloquine when given alone or in combination with sulphadoxine and pyrimethamine in Thai male and female subjects. *Eur J Clin Pharmacol*, **32**, 173–177.
14. Franssen G, Rouveix B, Le Bras J, Bauchet J, Verdier F, Michon C, Bricaire F (1989). Divided dose kinetics of mefloquine in man. *Br J Clin Pharmacol*, **28**, 179–184.
15. Mimica I, Fry W, Eckert G, Schwartz DE (1983). Multiple dose kinetic study of mefloquine in healthy male volunteers. *Chemotherapy*, **29**, 184–187.
16. Weidekamm E, Schwartz DE, Dubach UC, Weber B (1987). Single-dose investigation of possible interactions between the components of the antimalarial combination Fansimef™. *Chemotherapy*, **33**, 259–265.
17. Boudreau EF, Fleckenstein L, Pang LW, Childs GE, Schroeder AC, Ratnaratorn B, Phintuyothin P (1990). Mefloquine kinetics in cured and recrudescent patients with acute *falciparum* malaria and in healthy volunteers. *Clin Pharmacol Ther*, **48**, 399–409.
18. Wang NS, Guo XB, Liu QD, Fu LC, Li GQ, Arnold K (1990). Pharmacokinetics of the combination of pyrimethamine with sulfadoxine and mefloquine (FANSIMEF) in Chinese volunteers and the relative bioavailability of a lacquered tablet. *Chemotherapy*, **36**, 177–184.
19. Nosten F, ter Kuile F, Chongsuphajaisiddhi T, Na Bangchang K, Karbwang J, White NJ (1991). Mefloquine pharmacokinetics and resistance in children with acute *falciparum* malaria. *Br J Clin Pharmacol*, **31**, 556–559.
20. Hellgren U, Kihamia CM, Bergqvist Y, Rombo L (1991). Standard and reduced doses of mefloquine for treatment of *Plasmodium falciparum* in Tanzania – whole blood concentrations in relation to adverse reactions, *in vivo* response and *in vitro* susceptibility. *Am J Trop Med Hyg*, **45**, 254–262.
21. Mu JY, Israili ZH, Dayton PG (1975). Studies of the disposition and metabolism of mefloquine HCl (WR 142,490), a quinolinemethanol antimalarial, in the rat. Limited studies with an analogue, WR 30,090. *Drug Metab Disp*, **3**, 198–210.
22. Chevli R, Fitch CD (1982). The antimalarial drug mefloquine binds to membrane phospholipids. *Antimicrob Agents Chemother*, **21**, 581–586.
23. Chanthavanich P, Looareesuwan S, White NJ, Warrell DA, Warrell MJ, DiGiovanni JH, von Bredow J (1985). Intragastric mefloquine is absorbed rapidly in patients with cerebral malaria. *Am J Trop Med Hyg*, **34**, 1028–1036.
24. Jauch R, Griesser E, Oesterhelt G (1980). Metabolismus von Ro 21-5998 (Mefloquine) bei der Ratte. *Arzneimittelforschung/Drug Research*, **30**, 60–67.
25. Håkansson A, Landberg-Lindgren A, Björkman A (1990). Comparison of the activity *in vitro* of mefloquine and two metabolites against *Plasmodium falciparum*. *Trans R Soc Trop Med Hyg*, **84**, 503–504.
26. Schwartz DE, Eckert G, Ekue JMK (1987). Urinary excretion of mefloquine and some of its metabolites in African volunteers at steady state. *Chemotherapy*, **33**, 305–308.
27. Bygbjerg IC, Schapira A, Flachs H, Gomme G, Jepsen S (1983). Mefloquine resistance of *falciparum* malaria from Tanzania enhanced by treatment. *Lancet*, **ii**, 774–775.
28. Boudreau EF, Webster HK, Pavanand K, Thosingha L (1982). Type II mefloquine resistance in Thailand. *Lancet*, **II**, 1335.
29. Trenholme GM, Williams RL, Desjardins RE, Frischer H, Carson P, Rieckmann KH (1975). Mefloquine (WR 142,490) in the treatment of human malaria. *Science*, **190**, 792–794.
30. *Advances in Malaria Chemotherapy*. Technical Report Series No. 711 (1984). (Geneva: World Health Organization).
31. Kofi Ekue JM, Ulrich AM, Rwabwogo-Atenyi J, Sheth UK (1983). A double blind comparative trial of mefloquine and chloroquine in symptomatic *falciparum* malaria. *Bull World Health Organ*, **61**, 713–718.
32. Slutsker LM, Khoromana CO, Payne D, Allen CR, Wirima JJ. Heymann DL, Patchen L, Stekete RW (1990). Mefloquine therapy for *Plasmodium falciparum* malaria in children under 5 years of age in Malawi: *in vivo/in vitro* efficacy and correlation of drug concentration with parasitological outcome. *Bull World Health Organ*, **68**, 53–59.

33. Sowunmi A, Salako LA, Walker O, Ogundahunsi OAT (1990). Clinical trials of mefloquine in children suffering from chloroquine resistant *P. falciparum* malaria in Nigeria. *Trans R Soc Trop Med Hyg*, **84**, 761–764.
34. Oduola AMJ, Moyou-Somo RS, Kyle DE, Martin SK, Gerena L, Milhous WK (1989). Chloroquine resistant *Plasmodium falciparum* in indigenous residents of Cameroon. *Trans R Soc Trop Med Hyg*, **83**, 308–310.
35. Nosten F, ter Kuile F, Chongsuphajaisiddhi T, Luxemburger C, Webster HK, Edstein M, Phaipun L, Thew KL, White NJ (1991). Mefloquine-resistant *falciparum* malaria on the Thai–Burmese border. *Lancet*, **337**, 1140–1143.
36. Lobel HO, Bernard KW, Williams SL, Hightower A.W, Patchen LC, Campbell CC (1991). Effectiveness and tolerance of long-term prophylaxis with mefloquine. Need for a better dosing regimen. *JAMA*, **265**, 361–364.
37. Clyde DF, McCarthy VC, Miller RM, Hornick RB (1976). Suppressive activity of mefloquine in sporozoite-induced human malaria. *Antimicrob Agents Chemother*, **9**, 384–386.
38. Rieckmann KH, Trenholme GM, Williams RL, Carson PE, Frischer H, Desjardins RE (1974). Prophylactic activity of mefloquine hydrochloride (WR 142–490) in drug-resistant malaria. *Bull World Health Organ*, **51**, 375–377.
39. Webster LT (1990). Drugs used in the Chemotherapy of protozoal infections. In: *The Pharmacological Basis of Therapeutics*, 8th edn edited by A.G. Gilman, T.W. Rall, A.S. Nies and P. Taylor. (New York: Collier Macmillan) 988.
40. Collingnon P, Hehir, J, Mitchell D (1989). Successful treatment of *falciparum* malaria in pregnancy with mefloquine. *Lancet*, **i**, 967.
41. Nosten F, Karbwang J, White NJ, Honeymoon Na, Bangchang K, Bunnag D, Harinasuta T (1990). Mefloquine antimalarial prophylaxis in pregnancy: dose finding and pharmacokinetic study. *Br J Clin Pharmacol*, **30**, 79–85.
42. Edstein MD, Veenendal JR, Hyslop R (1988). Excretion of mefloquine in human breast milk. *Chemotherapy*, **34**, 165–169.
43. Karbwang J, Na Bangchang K, Bunnag D, Harinasuta T (1991). Pharmacokinetics and pharmacodynamics of mefloquine in Thai patients with acute *falciparum* malaria. *Bull World Health Organ*, **69**, 207–212.
44. Patchen LC, Campell CC, Williams SB (1989). Neurological reactions after a therapeutic dose of mefloquine. *N Engl J Med*, **321**, 1415.
45. Rouveix B, Bricaire F, Michon C, Franssen G, Le Bras J, Bernard J, Ajana F, Vienne JL (1989). Mefloquine and an acute brain syndrome. *Ann Intern Med*, **110**, 577–578.
46. Stuvier PC, Ligthelm RJ, Goud Th JLM (1989). Acute psychosis after mefloquine. *Lancet*, **ii**, 282.
47. World Health Organization. Review of central nervous system adverse events related to antimalarial drug, mefloquine (1985–1990) (WHO/MAL911063). (Geneva: World Health Organization).
48. Bem JL, Kerr L, Stürchler D (1992). Mefloquine prophylaxis: an overview of spontaneous reports of severe psychiatric reactions and convulsions. *J Trop Med Hyg*, **95**, 167–179.
49. Mefloquine. *Therapeutic Drugs*, edited by Sir Colin Dollery (1991) (London: Churchill Livingstone), pp. M35–M39.
50. *Martindale: The Extra Pharmacopoeia*, 30th edn, (1993) (London: Pharmaceutical Press), pp. 402–403.
51. Steffen R, Heusser R, Mächler R, Bruppacher R, Naef U, Chen D, Hofmann AM, Somaini B (1990). Malaria chemoprophylaxis among European tourists in tropical Africa: use, adverse reactions and efficacy. *Bull World Health Organ*, **68**, 313–322.
52. Lobel HO, Miani M, Eng T, Bernard KW, Hightower AW, Campbell CC (1993). Long-term prophylaxis with weekly mefloquine. *Lancet*, **341**, 848–851.
53. Stürchler D, Handschin J, Kaiser D, Kerr L, Mittelholzer M-L, Reber R, Fernex M (1990). Neuropsychiatric side effects of mefloquine. *N Engl J Med*, **322**, 1752–1753.
54. Weinke T, Trautmann M, Held T, Weber G, Eichenlaub D, Fleischer K, Kern W, Pohle HD (1991). Neuropsychiatric side effects after the use of mefloquine. *Am J Trop Med Hyg*, **45**, 86–91.
55. *WHO Model Prescribing Information*. Drugs used in parasitic diseases (1990), (Geneva: World Health Organization).
56. *Dictionnaire Vidal*, 70th edn, (1994), P795.
57. Rombo L, Angel VH, Friman G, Hellgren U, Mittelholzer ML, Stürchler D (1993). Comparative tolerability and kinetics during long-term intake of Lariam® and Fansidar® for malaria prophylaxis in nonimmune volunteers. *Trop Med Parasitol*, **44**, 254–256.

Melarsoprol

Chemical structure

Melarsoprol

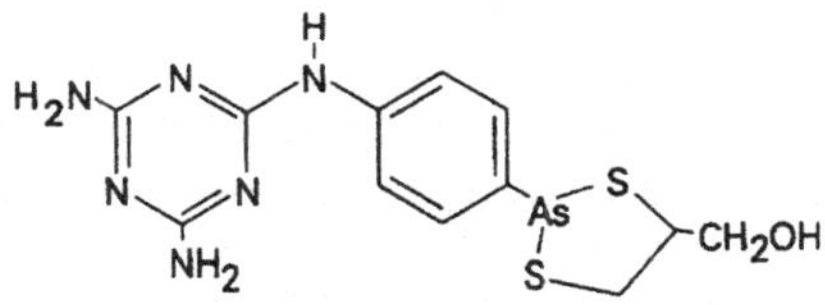

Physical properties

MW 398, pKa not known. Practically insoluble in water. Melarsoprol is marketed as a 3.6% solution in propylene glycol in 5 ml ampoules. The ampoules should be stored in the dark at temperatures below 25 °C.

Pharmacology and mechanism of action

Melarsoprol (Mel B) is a trivalent arsenical compound which was introduced into clinical medicine in 1949 by Friedman (1). It is active against all stages of *Trypanosoma brucei gambiense* and *Trypanosoma brucei rhodesiense* infections. However, because of toxicity, it is only used in late-stage trypanosomiasis.

The mechanism of action of melarsoprol is not well characterized. However, there is evidence showing that melarsoprol forms a complex with parasite trypanothione which protects the parasite from oxidant damage and lysis (2). The formation of trypanothione depends on polyamine biosynthesis which is blocked by another trypanosomicide, eflornithine. A possible synergistic effect of eflornithine and melarsoprol has been reported by Jennings (3).

Pharmacokinetics

A specific analytical method has not been described, but some pharmacokinetic data have been obtained using ELISA (4) and a bioassay method (5) of unknown specificity.

Melarsoprol is only given intravenously, since the solution is too irritant for intramuscular use. The drug passes the blood–brain barrier and may attain concentrations of 1–2% of those in plasma. The apparent volume of distribution is more than 100 l, and the terminal plasma elimination half-life is about 35 hours (6). The drug or its metabolites are eliminated primarily through the bile and to a lesser extent by the kidneys (7, 8).

Clinical trials

In an open clinical study Robertsson (9) treated 71 patients (7 infected with *Trypanosoma brucei gambiense* and 64 with *Trypansoma brucei rhodesiense*) with melarsoprol (total dose of 1260 mg spread over 21 days) and reported a cure rate of 90% after 3–6 months of follow-up. Ferreira *et al.* (10) treated 335 patients with trypanosomiasis with 3.6 mg/kg of melarsoprol (max. 200 mg) every second day. A substantial number of the patients (84%) had signs of CNS involvement, i.e., raised CSF protein and an increased number of white cells in CSF. Some of the patients were pre-treated with tryparsamide, suramin or pentamidine. The patients were followed up from 6 months to 10 years. Cure rates after 6 months to 10 years were 74–85% and 86–96%, with pre-treatment and without, respectively. Death rates were 5–7% and 4–9%, respectively for the two groups. Patients with earlier stages of the disease tended to recover earlier than those with the severe type of the disease.

Dutertre and Labusquière (11) treated trypanosomiasis with a lower total melarsoprol dose than is commonly used. A total of 670 patients without CNS involvement were given 1 or 3 injections of melarsoprol of 3.6 mg/kg per injection. Another group of 1318 patients with signs of CNS involvement were given 1 or 2 treatment schedules with 3 injections of melarsoprol (3.6 mg/kg). The cure rates were 93% in the first group and 88% in the second group of patients. The death rate was 0.6% and 5%, respectively. The type and severity of the side effects were not discussed.

In an open prospective study Wellde *et al.* (12) treated 113 patients with *Trypanosoma brucei rhodesiense* with standard doses of melarsoprol. The patients were followed up for 3 years from 1981 until 1984. During the course of the treatment 14 patients died (5%). After the completion of the follow-up period 92% were cured, 1.4% relapsed, and a further 6% died because of unknown causes.

Indications

Melarsoprol is only indicated for the treatment of late stage (CNS involvement) of *Trypanosoma brucei gambiense* and *Trypanosoma brucei rhodesiense*. It is a clinical experience that patients with *Trypanosoma brucei rhodesiense* who relapse usually respond to a second course of melarsoprol, while those with *Trypanosoma brucei gambiense* who relapse rarely do so (13). Thus, patients with *Trypanosoma brucei gambiense* who do not respond to the first treatment course of melarsoprol should be switched to eflornithine.

Pregnancy and lactation

Teratogenicity in animals is unknown. Documentation in man is still limited. Pépin and Milord (13), have reported to have administered melarsoprol to several pregnant women without any ill effects to the new-born. In another report, Lowenthal (14), has treated a single woman in her twenty-first week of pregnancy with suramin and melarsoprol for an advanced case of cerebral trypanosomiasis. The woman delivered a healthy child. Because of the severity of the disease, melarsoprol should not be withheld from a pregnant woman with trypanosomiasis.

Its excretion into breast milk is unknown.

Side effects

The most serious side effect with the use of melarsoprol is encephalopathy which is usually seen between 5 and 12 days after the first dose. It occurs in about 2–10% of the patients of which 50–75% of them may die (15). Melarsoprol encephalopathy has been classified as two different entities: reactive and haemorrhagic encephalopathies. Reactive encephalopathy is relatively more common and is characterized by mental and motor excitation, drowsiness which progresses into coma and convulsions. It is often reversible. The haemorrhagic type has been described as a rare entity, which is nearly always fatal. While reactive encephalopathy is attributed to drug-related immunological response, the haemorrhagic encephalopathy is thought to be due to melarsoprol toxicity (15). Other authors have disputed this distinction and have described the two types of encephalopathy merely as various stages of severity of the same condition (13). Although the exact mechanism of melarsoprol encephalopathy is as yet unknown and remains controversial, recent reports say that it is likely to be due to an immunological reaction triggered by high initial doses of melarsoprol (16, 17).

Management of these reactions includes corticosteroids and hyper-osmotic solutions to reduce cerebral oedema, anticonvulsants and subcutaneous injections of adrenaline. The administration of corticosteroids during melarsoprol treatment may reduce the risk of encephalopathy. In a recent randomized comparative study in *Trypanosoma brucei gambiense*, the incidence of encephalopathy was only 12% in patients treated with prednisolone (1 mg/kg daily to a maximum of 40 mg) together with melarsoprol as compared to those treated with melarsoprol alone who had an incidence of 35% (18).

Other reactions such as albuminuria and abdominal colic are common (10–15% incidence). After the first administration of melarsoprol, as with other antitrypanosomal drugs, a fever of up to 40 °C is seen in about 60% of the patients (19). Nausea, vomiting and diarrhoea may be seen. Skin reactions, arthralgia, agranulocytosis, aplastic anaemia, thrombocytopenia, renal and hepatic failure and Guillain–Barré-like syndrome have been reported occasionally (9–11, 15, 20).

Contraindications and precautions

Patients should be hospitalized and well supervised during melarsoprol treatment. Patients in poor general condition may not tolerate the drug. The drug can evoke severe haemolytic reactions in patients with glucose 6-phosphate dehydrogenase (G6PD) deficiency. The administration of melarsoprol to leprous patients may induce erythema nodosum. The drug may aggravate the condition of the patient during viral infections such as influenza. In such situations treatment may be postponed. The solution used for i.v. administration contains propylene glycol which is highly irritant to the tissues. Extravascular leakage will cause severe tissue destruction and thrombophlebitis. Therefore, the solution must be carefully and slowly injected with a fine needle. Melarsoprol should not be given to patients with the early stage of the disease, since the drug can cause encephalopathy even in such patients (13).

Dosage

The treatment regimens and duration of therapy of melarsoprol vary between different endemic areas. In some centres progressively increasing doses are used, while maximum daily doses are practised in others. The use of pentamidine and suramin before melarsoprol administration vary also in different centres. Although the usefulness of corticosteroids has been shown, its use is not uniform among different countries.

The WHO recommended dosage regimen practised in some areas (21) is given below. The treatment schedules used at present are unpractical: it is possible that shorter treatment courses may be as effective. Data supporting shorter courses are, however, lacking.

Adults and children

Day	*Drug*	*Dose* (mg/kg)
For *T.b. rhodesiense* infection, as used in Kenya and Zambia		
1	suramin	5.0
3	suramin	10.0
5	suramin	20.0
7	melarsoprol	0.36
8	melarsoprol	0.72
9	melarsoprol	1.10
16	melarsoprol	1.40
17	melarsoprol	1.80
18	melarsoprol	1.80
25	melarsoprol	2.20
26	melarsoprol	2.90
27	melarsoprol	3.60
34	melarsoprol	3.60
35	melarsoprol	3.60
36	melarsoprol	3.60
For *T.b. rhodesiense* infection, as used in Uganda and the United Republic of Tanzania		
1	suramin	5.0
3	suramin	10.0
5	melarsoprol	1.80
6	melarsoprol	2.20
7	melarsoprol	2.56
14	melarsoprol	2.56
15	melarsoprol	2.90
16	melarsoprol	3.26
23	melarsoprol	3.60
24	melarsoprol	3.60
25	melarsoprol	3.60
For *T.b. gambiense* infection, as used in Côte d'Ivoire		
1	pentamidine i.m.	4.0
2	pentamidine i.m.	4.0
4	suramin	1.20
5	melarsoprol	2.40
6	melarsoprol	3.60
17	melarsoprol	1.20
18	melarsoprol	2.40
19	melarsoprol	3.60
20	melarsoprol	3.60
30	melarsoprol	1.20
31	melarsoprol	2.40

Day	*Drug*	*Dose* (mg/kg)
32	melarsoprol	3.60
33	melarsoprol	3.60
34	melarsoprol	3.60
35	melarsoprol	3.60
36	melarsoprol	3.60

Preparations

- Arsobal® (Specia). 5 ml ampoules of 36 mg/ml in propylene glycol solution.

References

1. Friedman EAH (1949). Mel B in the treatment of human trypanosomiasis. *Amer J Trop Med Hyg*, **29**, 173–180.
2. Fairlamb AH, Henderson GB, Cerami A (1989). Trypanothione is the primary target for arsenical drugs against African trypanosomes. *Proc Natl Acad Sci USA*, **86**, 2607–2611.
3. Jennings FW (1988). Chemotherapy of trypanosomiasis: the potentiation of melarsoprol by concurrent difluoromethylornithine (DFMO) treatment. *Trans R Soc Trop Med Hyg*, **82**, 572–573.
4. Maes L, Vanderveken M, Hamers R, Doua F, Cattand P (1988). The monitoring of trypanocidal treatment with a sensitive ELISA method for measuring melarsoprol levels in serum and in cerebrospinal fluids. *Ann Soc Belge Méd Trop*, **68**, 219–231.
5. Burri C, Brun R (1992). An *in vitro* bioassay for quantification of melarsoprol in serum and cerebrospinal fluids. *Trop Med Parasitol*, **4**, 223–225.
6. Burri C, Baltz T, Girond C, Doua, F, Welker HA, Brun R (1993). Pharmacokinetic properties of the trypanosomicidal drug melarsoprol. *Pharmacol*, **39**, 225–234.
7. Cristau B, Placidi M, Legait JP (1975). Etude de l'excrétion de l'arsenic chez le trypanosomé traité au mélarsoprol (Arsobal). *Méd. Trop*, **35**, 389–401.
8. Hawking F (1962). Estimation of the concentration of melarsoprol (Mel B) and Mel W in biological fluids by bioassay with trypanosomes *in vitro*. *Trans R Soc Trop Med Hyg*, **56**, 354–363.
9. Robertsson DHH (1963). The treatment of sleeping sickness (mainly due to *Trypanosoma rhodesiense*) with melarsoprol. II. An assessment of its curative value. *Trans R Soc Trop Med Hyg*, **57**, 176–183.
10. Ferreira FSC, Costa FMC (1963). Restatos do tratamento da tripanosomiose humana africana com o arsobal. *Gaz Med Portoguesa*, **166**, 11–618.
11. Dutertre J, Labusquière R. (1966) La thérapeutique de la trypanosomiase. *Med Trop*, **26**, 342–356.
12. Wellde BT, Chumo DA, Reardon MJ (1989). Treatment of rhodesian sleeping sickness in Kenya. *Ann Trop Med Parasitol*, **1**, 99–109.
13. Pepin J, Milord F (1994). The treatment of human African trypanosomiasis. *Adv Parasitol*, **33**, 1–47.
14. Lowenthal MN (1971). Trypanosomiasis successfully treated in a pregnant woman. *Med J Zambia*, **5**, 175.
15. Robertsson DHH (1963). The treatment of sleeping sickness (mainly due to *Trypanosoma rhodesiense*) with melarsoprol. I. Reactions observed during treatment. *Trans R Soc Trop Med Hyg*, **57**, 1246–1250.
16. Haller L, Adams H, Merouze F, Dago A (1986). Clinical and pathological aspects of human African trypanosomiasis (*T.b. gambiense*) with particular reference to reactive encephalopathy. *Am J Trop Med Hyg*, **35**, 94–99.
17. Pepin J, Milord F (1991). African trypanosomiasis and drug-induced encephalopathy: risk factors and pathogenesis. *Trans R Soc Trop Med Hyg*, **85**, 222–224.
18. Pepin J, Milord F, Guern C, Mpia B, Ethier L, Mansinsa D (1989). Trial of prednisolone for prevention of melarsoprol-induced encephalopathy in gambiense sleeping sickness. *Lancet*, **i**, 1246–1250.

19. Whittle HC, Pope HM (1972). The febrile response to treatment in Gambian sleeping sickness. *Ann Trop Med Parasitol*, **66**, 7–14.
20. Gherardi RK, Chariot P, Vanderstigel M, Malapert D, Verroust J, Astier A, Brun-Buisson C, Schaeffer A (1990). Organic arsenic-induced Guillain–Barré-like syndrome due to melarsoprol: a clinical, electrophysiological, and pathological study. *Muscle & Nerve*, **13**, 637–645.
21. *WHO Model Prescribing Information*. Drugs used in parasitic diseases (1990). (Geneva: World Health Organization).

Metrifonate

Chemical structure

Metrifonate

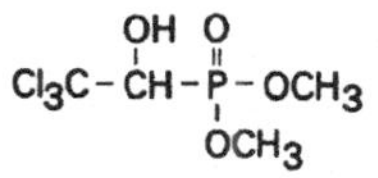

Dichlorvos

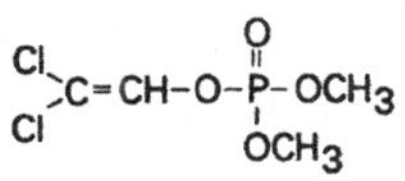

Physical properties

MW 257. Neutral compound. 1 g dissolves in 7 ml of water. Decomposes to dichlorvos in alkaline solutions. Stable below pH 5.

Pharmacology and mechanism of action

Metrifonate is an organophosphorus compound, first introduced as an insecticide in 1952 and a little later as an anthelminthic. Early clinical studies have reported it to be effective against a wide number of helminthic infections including schistosomiasis, ascariasis, ancylostomiasis, and trichuriasis (1). The drug has been tried in onchocerciasis with limited success (2). It is also an experimental drug in Alzheimer's disease (3). Today, it is mainly used against *Schistosoma (S) haematobium.*

The mechanism of action of metrifonate is unknown. The only pharmacological action described hitherto is its inhibitory effect on cholinesterases, which is due to its rearrangement product, dichlorvos. Dichlorvos, as a drug is used widely in veterinary medicine and has been given to man as a slow release preparation (4, 5). *In vitro*, metrifonate paralyses both *S. haematobium* and *S. mansoni* (6). However, clinically it is effective only against *S. haematobium*. Although this paradox has been explained to be due to the different locations of the two worms in man (7, 8), recent reports suggest that *S. haematobium* may be more sensitive to metrifonate than *S. mansoni* because of much higher levels of cholinesterase activity in its tegument (9).

Pharmacokinetics

Specific GC (10, 11, 12) and GC/MS (11,12) methods have been described for the determination of metrifonate and its non-enzymatic product, dichlorvos, respectively. A new HPLC method with UV detection (13) has also been described recently for both compounds, but its reliability has been questioned (14).

Absolute oral availability is unknown. After oral intake metrifonate is well absorbed, and peak plasma levels are reached after 1–2 hours (15, 16). In the human body it is transformed into an active compound, dichlorvos (DDVP) by a non-enzymatic process (10). This conversion occurs also *in vitro* and is highly pH dependent (10, 16). The whole blood concentration of dichlorvos is at all times about 1% of that of the parent drug (12, 15, 16).

Thus, metrifonate has been described to act as a slow release preparation of dichlorvos (17). In rats, both metrifonate and dichlorvos are degraded largely by hydrolysis with the excretion of large quantities of dimethyl phosphate. O-demethylation of both compounds also occurs at a minor rate. In man, the metabolism is likely to be similar to that reported in animals (17). The plasma elimination half-life of metrifonate in man is around 3 hours (12, 15, 16). Elimination is via the kidneys, mainly in the form of glucuronidates (17).

Clinical trials

Metrifonate was first tested as a human anthelminthic drug in 1960 (18). Early clinical and toxicological studies conducted in Zaire showed metrifonate to be safe with doses between 7 and 24 mg/kg once daily for 2 days (18, 19). The only effect seen was a decrease of blood cholinesterases. Cerf *et al.* in 1962 (20), gave the drug to more than 2000 patients with ascariasis, ancylostomiasis, bilharziasis (*S. mansoni*), trichuriasis, creeping eruptions, taeniasis and strongyloidiasis. The doses used were 7.5–20 mg/kg daily for 2 days. The drug had cure rates of 60–100% for most parasites including *S. mansoni*. Similar studies carried out in Egypt (21) further confirmed the safety of the drug but demonstrated no significant effect against helminths or *S. mansoni*, but reported good efficacy against *S. haematobium*. A dose of 5–10 mg/kg repeated once daily for 6–12 days was used. Most of these early studies were not standardized and they used impure preparations of the drug.

During the late 1960s and early 1970s, the World Health Organization got involved and more carefully conducted trials were carried out (22, 23). These studies further confirmed the fall in blood cholinesterases after metrifonate administration and established the presently used dosage regimen of 3 doses of 7.5 mg/kg given at fortnightly intervals against *S. haematobium*. The studies also found no relationship between the degree of cholinesterase inhibition and the occurrence of side effects.

Today, metrifonate is used only against *S. haematobium*. Single oral doses of 7.5–10 mg/kg given 3 times at fortnightly intervals give cure rates of 60–70% and egg reduction rates of up to 90% 4 weeks after the last dose (22, 23). More recent studies have also reported that metrifonate is effective against ancylostomiasis (24). A number of studies have reported the drug to improve haemoglobin levels of treated children due to its effect on ancylostomiasis (25, 26). Because of the difficulties associated with the standard regimen of metrifonate against schistosomiasis, most areas have used single dosage regimens repeated every 6 months or yearly. Single dose regimens of metrifonate seem, however, to be ineffective (25, 27, 28). In recent clinical studies in Somalia, a simplified dosage regimen of 5 mg/kg given 3 times in 1 day have been reported to be as effective and safe as the standard dosage regimen (29, 30). Although this regimen might still have compliance problems, it is a major advance over the standard regimen.

Indications

For the treatment of *Schistosoma haematobium* infections. During mass treatment programmes, when cost is a major factor, metrifonate may be preferred over praziquantel. When radical treatment is desired and cost is not a problem, praziquantel is the first drug of choice, i.e. in places where re-infection is not expected.

Pregnancy and lactation

Metrifonate is not embryotoxic or carcinogenic in laboratory animals (31), however, it is reported to have caused cerebral hypoplasia in piglets (32). Experience in humans is limited. An infant born with massive hydrocephalus and a large meningomyelocele whose mother had been treated twice during the second month of her pregnancy with metrifonate has been reported (33). The report is old and the association was probably co-incidental. Treatment with metrifonate should be postponed until delivery.

Excretion into breast milk is unknown.

Side effects

Despite extensive toxicological and clinical studies no major side effects have been observed with the recommended dose (17, 22). One of the most important side effects of the drug, is its effect on blood cholinesterases. Soon after its intake, both plasma and erythrocyte cholinesterase levels are inhibited to zero and to 80%, respectively. Normal plasma cholinesterase levels return after 4 weeks, but it takes longer for the recovery of the erythrocyte cholinesterase (23). Although no correlation seems to exist between the dose and the degree of cholinesterase inhibition, a good relationship was found between the occurrence of side effects and the plasma levels of the drug (34). Side effects commonly reported include nausea, vomiting, headache, abdominal pain, vertigo, and fatigue. They are low in frequency and severity and they usually disappear spontaneously within a few hours after drug intake.

In the case of metrifonate intoxication, pralidoxime iodide is used as an antidote (1 g is injected intravenously in 2 minutes. The dose may be repeated after 20 minutes if symptoms persist). In addition, atropine should be given in high doses, i.e. 2–4 mg i.v. every 3–10 minutes to a maximum daily dose of 50 mg.

Contraindications

Metrifonate should not be given to patients taking suxamthonium. In areas where organophosphorus insecticides have been sprayed, the community may already have low levels of blood cholinesterases. Special precautions are needed in such situations.

Interactions

Metrifonate prolongs the muscle-relaxing effect of succinylcholine.

Dosage

An optimal dosage regimen of metrifonate is not yet available. There is a need for further studies to determine the best regimen. The standard regimen (7.5 mg/kg on 3 occasions at 2-week intervals) recommended by the WHO (35) is complicated and in most areas a single dosage regimen of 10 mg/kg is used which is sometimes repeated at different intervals. A simpler one day treatment regimen has been published (30), and is worth testing in different endemic areas.

Preparations

- Bilarcil® (Bayer). Tablets 100 mg.

References

1. Cerf J, Lebrun A, Dierichx J (1962). A new approach to helminthiasis control: The use of an organophosphorus compound. *Am J Trop Med Hyg*, **11**, 514–517.

2. Awadzi K, Gilles HM (1980). The chemotherapy of onchocerciasis, III: a comparative study of diethylcarbamazine DEC and metrifonate. *Ann Trop Med Parasitol*, **74**, 210–217.
3. Moriearty PL, Womack CL, Dick BW, Colliver JA, Robbs RS, Becker RE (1991). Stability of peripheral hematological parameters after chronic acetylcholinesterase inhibition in man. *Am J Hematol*, **37**, 280–282.
4. Cervoni WA, Oliver-Gonzalez J, Kaye S, Slomka MB (1969). Dichlorvos as a single-dose intestinal anthelminthic therapy for man. *Am J Trop Med Hyg*, **18**, 912–919.
5. Chavarria APA, Swartzwelder JC, Villarejos VM, Kotcher E, Arguedas J (1969). Dichlorvos, an effective broad spectrum anthelminthic. *Am J Trop Med Hyg*, **18**, 907–911.
6. Bueding E, Liu CL, Rogers SH (1972). Inhibition by metrifonate and dichlorvos of cholinesterases in schistosomes. *Br J Pharmacol*, **46**, 480–487.
7. Forsyth DM, Rashid C (1967). Treatment of urinary schistosomiasis with trichlorofon. *Lancet*, **ii**, 909–912.
8. Feldmeier H, Doehring E, Daffalla AA, Omer AHS, Dietrich M (1982). Efficacy of metrifonate in urinary schistosomiasis: Comparison of reduction of *Schistosoma haematobium* and *S. mansoni* eggs. *Am J Trop Med Hyg*, **31**, 1188–1194.
9. Camacho M, Tarrab-Hazdai R, Espinoza B, Arnon R, Agnew A (1994). The amount of acetylcholinesterase on the parasite surface reflects the differential sensitivity of schistosome species to metrifonate. *Parasitology*, **108**, 153–160.
10. Nordgren I, Bergström M, Holmstedt B, Sandoz M (1978). Transformation and action of metrifonate. *Arch Toxicol*, **41**, 31–41.
11. Ameno K, Fuke C, Ameno S, Kiriu T, Ijiri I (1989). A rapid and sensitive quantitation of Diptrex in serum by solid-phase extraction and gas chromatography with flame thermionic detection. *J Anal Toxicol*, **13**, 150–151.
12. Villén T, Aden Abdi Y, Ericsson Ö, Gustafsson LL, Sjöqvist F (1990). Determination of metrifonate and dichlorvos in whole blood using gas chromatography and gas chromatography-mass spectrometry. *J Chromatogr*, **529**, 309–317.
13. Unni LK, Hannant ME, Becker RE (1992). High performance liquid chromatographic method using ultraviolet detection for measuring metrifonate and dichlorvos levels in human plasma. *J Chromatogr*, **573**, 99–103.
14. Aden Abdi Y, Villén T, Gustafsson LL, Ericsson Ö, Sjöqvist F (1993). Methodological commentary on the analysis of metrifonate and dichlorvos in biological samples. *J Chromatogr*, **612**, 336–337.
15. Nordgren I, Bengtsson E, Holmstedt B, Pettersson BM (1981). Levels of metrifonate and dichlorvos in plasma and erythrocytes during treatment of schistosomiasis with Bilarcil. *Acta Pharmacol Toxicol*, **49**, 79–86.
16. Aden Abdi Y, Villén T (1991). Pharmacokinetics of metrifonate and its rearrangement product in whole blood. *Pharmacol Toxicol*, **68**, 137–139.
17. Holmstedt B, Nordgren I, Sandoz M, Sundwall A (1978). Metrifonate: Summary of toxicological and pharmacological information available. *Arch Toxicol*, **41**, 3–29.
18. Lebrun A, Cerf C (1960). Note preliminaire sur la toxicité pour l'homme d'un insecticide organophosphore (Diptrex). *Bull WHO*, **22**, 579–582.
19. Beheyet P, Lebrun A, Cerf J, Dierichx J, Degroote V (1961). Étude de la toxicité pour l'homme d'un insecticide organophosphore. *Bull WHO*, **24**, 465–475.
20. Cerf J, Lebrun A, Dierichx J (1962). A new approach to helminthiasis control: The use of an organophosphorous compound. *Am J Trop Med Hyg*, **11**, 514–517.
21. Abdalla A, Saif M, Taha A, Ashmawy H, Tawfik J, Abdel-Fattah F, Sabet S, Abdel-Meguid M (1965). Evaluation of an organophosphorous compound, Diptrex, in the treatment of Bilharziasis. *J Egypt Med Assoc*, **48**, 262–273.
22. Davis A, Bailey DR (1969). Metrifonate in urinary schistosomiasis. *Bull WHO*, **41**, 209–224.
23. Plestina R, Davis A, Bailey DR (1972). Effect of metrifonate on blood cholinesterases in children during the treatment of schistosomiasis. *Bull WHO*, **46**, 747–759.
24. Kurz KM, Stephenson LS, Latham MC, Kinoti S (1986). The effectiveness of metrifonate in reducing hookworm infection in Kenyan school children. *Am J Trop Med Hyg*, **35**, 571–574.
25. Stephenson LS, Kinoti SN, Latham MC, Kurz K, Kyobe J (1989). Single dose metrifonate or praziquantel treatment in Kenyan Children. I. Effects on *Schistosoma haematobium*, hookworm, haemoglobin levels, splenomegaly, and hepatomegaly. *Am J Trop Med Hyg*, **41**, 445–453.
26. Stephenson LS, Latham MC, Kurz KM, Kinoti SN, Oduori ML, Crompton DW (1985).

Relationship of *Schistosoma haematobium*, hookworm and malarial infections and metrifonate treatment to hemoglobin level in Kenya school children. *Am J Trop Med Hyg*, **341**, 519–528.
27. Pugh RN, Teesdale CH (1983). Single dose oral treatment in urinary schistosomiasis: a double blind trial. *BMJ*, **286**, 429–432.
28. Tswana JA, Mason PR (1985). Eighteen-month follow up on the treatment of urinary schistosomiasis with a single dose of metrifonate. *Am J Trop Med Hyg*, **34**, 746–749.
29. Aden Abdi Y, Gustafsson LL, Elmi SA (1987). A simplified dosage schedule of metrifonate in the treatment of *Schistosoma haematobium* infection in Somalia. *Eur J Clin Pharmacol*, **32**, 437–441.
30. Aden Abdi Y, Gustafsson LL (1989). Field trial of the efficacy of a simplified and standard metrifonate treatments of *Schistosoma haematobium*. *Eur J Clin Pharmacol*, **37**, 371–374.
31. Machemer L (1981). Chronic toxicity of metrifonate. *Acta Pharmacol Toxicol*, **49**, 15–28.
32. Knox B, Askaa J, Basse A, Bitsch V, Eskildsen M, Mandrup M, Ottosen HE, Overby E, Pedersen KB, Rasmussen F (1978). Congenital ataxia and tremor with cerebral hypoplasia in piglets borne by sows treated with neguvon (metrifonate) during pregnancy. *Nor Vet Med*, **30**, 535–545.
33. Monson MH, Alexander K (1984). Metrifonate in pregnancy. *Trans R Soc Trop Med Hyg*, **78**, 565.
34. Aden Abdi Y, Villén T, Ericsson Ö, Gustafsson LL, Dahl-Puustinen M-L (1990). Metrifonate in healthy volunteers: interrelationship between pharmacokinetic properties, cholinesterase inhibition and side effects. *Bull WHO*, **68**, 731–736.
35. *WHO Model Prescribing Information. Drugs used in parasitic diseases* (1990), Geneva: World Health Organization).

Metronidazole

Chemical structure

Metronidazole

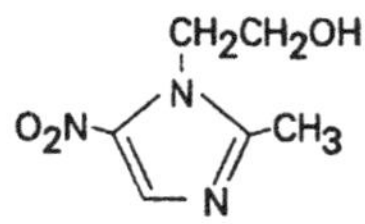

Physical properties

MW 171; pKa 2.6. 1 g dissolves in 100 ml of water. Protect from light.

Pharmacology and mechanism of action

Metronidazole is a 5-nitroimidazole derivative which was originally introduced against *Trichomonas vaginalis* in 1960. Soon it was shown to possess a broad spectrum of activity against other protozoal infections such as amoebiasis and giardiasis, and more recently against infections due to anaerobic bacteria (1).

The mechanism of action of metronidazole is not well understood. In the parasite, the 5-nitro group of the drug undergoes reductive transformation to a cytotoxic intermediate which binds to the helical structure of the DNA leading to strand breakage and eventual cell death (2).

Pharmacokinetics

Specific HPLC methods (3, 4) have been described for the determination of metronidazole and its metabolites.

Metronidazole can be administered orally, intravenously, rectally or intravaginally as a suppository or cream. Its bioavailability is complete after oral administration (close to 100%) and food does not significantly affect its oral absorption (5). The bioavailability of the drug after rectal administration is about 80% of an equal oral dose. However, it is slowly and poorly absorbed after intravaginal administration (5, 6). There is some data showing that the drug might have dose-dependent metabolism at doses of 2000 mg or above (7).

Following oral administration of a single 500 mg dose, average peak plasma levels of 13 μg /ml were reached after 2 hours (8). Less than 20% of the drug is bound to plasma proteins (5), and is thus widely distributed throughout the body with an apparent volume of distribution which varied between the studies from 0.6 to 1.1 l/kg (9). Concentrations approximating those in plasma can be obtained in various tissues including the CSF, bile, breast milk, and saliva (5). Metronidazole is metabolized by the liver to hydroxy- and acid metabolites which are excreted partially as glucuronides in the urine. The metabolites have some pharmacological activities (10, 11). The average plasma elimination half-life reported in several studies using HPLC methods was around 7 hours. The half-life of the hydroxy metabolite exceeds that of the parent drug, ranging from 9.5 to 19.2 hours in patients with normal kidney function (9). Over 70% of an oral dose is excreted through the kidneys over 48 hours mainly as the hydroxy (30–40%) and acid metabolites (10–22%). Less than 10% is excreted as the parent drug (12). Around 14% of the administered dose may be excreted with the faeces (13).

In patients with renal failure, the half-life of the parent drug is unchanged, while that of the hydroxy metabolite is prolonged 4- to 17-fold (14, 15). The metabolic clearance of metronidazole can be significantly affected by severe liver impairment (16). Metronidazole is effectively removed by haemodialysis, and dosage supplementation is necessary in such patients (17).

The disposition of metronidazole in children is similar to that in adults using a weight-related dose (18). However, longer elimination half-lives and reduced clearance of the drug have been reported in newborns (3-fold increase in half-lives) (19), and in severely malnourished children (2-fold increase in half-lives) (20).

Clinical trials

A large number of studies have been conducted to evaluate the efficacy of metronidazole in amoebiasis in comparison with tinidazole. Most of these studies were open trials where severity of the diseases and experimental designs varied.

In one of the early studies, Islam and Hasan (21) compared the efficacy of metronidazole (2 g per day) with that of tinidazole (2 g per day) in 16 patients suffering from amoebic liver abscess. They continued treatment until it was no longer necessary on clinical grounds and found that the duration of treatment required with tinidazole was 4 days compared to 7 days for metronidazole. In another study, Bakshi *et al.* (22) found that tinidazole (2 g per day for 2 days) produced a complete recovery in 96% of 48 patients with amoebic dysentery and giardiasis, while metronidazole (2 g per day for 2 days) cured only 76% of 49 children. The authors reported that the disappearance of symptoms was more rapid with the tinidazole group. Similar results have been reported by Kokhani *et al.* (23) who used similar dosage regimens of metronidazole and tinidazole. Simjee *et al.* (24) compared metronidazole ($n = 27$ patients) versus tinidazole ($n = 21$ patients) against amoebic liver abscess with single daily doses of 2 g per day for 5 days. All patients were cured with the exception of 6 patients (2 from the metronidazole group and 4 from the tinidazole group) who needed a second course of therapy. Side effects reported were generally mild and were gastrointestinal in nature and were more common with the metronidazole group.

In trials conducted to evaluate its efficacy in the treatment of giardiasis, metronidazole has been shown to be highly effective for both adults and children (above 90% cure rate). In a comparative clinical trial, 80 patients with *Giardia lamblia* infection were treated either with metronidazole (0.5 g daily for 10 days) or tinidazole (single 2 g dose) or ornidazole (single 1 g dose). After follow-up, cure rates were 95% for metronidazole, 90% for tinidazole and 97% for ornidazole. No side effects were reported (25). However, most comparative studies have reported tinidazole to be more effective than metronidazole when used as a single dose (26, 27).

In a double-blind study, metronidazole (400 mg three times daily for 10 days) was compared to placebo against *Dracunculus medinensis*. A cure rate of 85% was observed. Only 4% needed more than 20 days treatment and one case failed to respond. Metronidazole was determined to be significantly better than placebo and rapid improvement of symptoms occurred (28). Similar results have been reported by others (29, 30). Side effects reported in these studies were tolerable. Despite the claims, the role of chemotherapy in the control of dracunculiasis is not very clear.

Indications

Against infections caused by *Trichomonas vaginalis*, *Entamoeba histolytica* (acute intestinal type and liver abscesses), *Giardia lamblia* and *Dracunculus medinensis*. During treatment of trichomoniasis it is wise to treat the male partner as well. In amoebiasis, a luminal amoebicide is added to eliminate surviving organisms in the colon. Metronidazole is also used for the treatment of infections due to anaerobic bacteria.

Pregnancy and lactation

Teratogenicity has not been reported in rabbits, rats and mice (31). Metronidazole readily crosses the placental barrier attaining a cord maternal plasma ratio of approximately 1 (32). Although thousands of pregnant women have been given the drug during pregnancy, evidence of teratogenicity has never been reported. However, in some of the studies there has been an increased risk of teratogenicity particularly when the drug was used during the first trimester (33). Metronidazole should be avoided during early pregnancy unless the condition of the patient makes it necessary for its use.

Excretion into breast milk is rapid, and nursed babies may get plasma levels of about 15% of that of the mother (34). Nursing should be stopped temporarily during treatment with metronidazole.

Side effects

Side effects with doses used to treat protozoal infections are usually mild, reversible and self-limiting and may affect 4% to 5% of treated patients. The most common are gastrointestinal disturbances (nausea, vomiting, epigastric pain, metallic taste, furring of the tongue), intolerance to alcohol (disulfiram-like effect) and central nervous system effects (headache, dizziness and sleepiness) (9). Other side effects reported include urticaria, darkening of the urine with a reddish-brown discoloration and transient neutropenia (31). During prolonged high doses, the drug may cause severe neurotoxic side effects such as peripheral neuropathy, paraesthesia and epileptiform seizures (9, 31). Few case reports of bone marrow depression (34), gynecomastia (36) and acute pancreatitis (37) have been reported.

Although metronidazole is mutagenic in bacteria and carcinogenic in rodents, no association with human cancer has been proven (33).

Contraindications and precautions

Dosage reductions should be made in patients with severe hepatic failure. Because of its potential neurotoxicity and neutropenia the drug should be given with caution to patients with diseases of the CNS or with a history of blood dyscrasia. Patients should be warned of a disulfiram-like reaction if the drug is taken together with alcohol. Metronidazole should be used with extra caution in patients being treated with warfarin (see interactions).

Interactions

Metronidazole is a weak inhibitor of alcohol dehydrogenase. Simultaneous administration of metronidazole and disulfiram has been reported to cause an acute psychosis or mental confusion. This effect was observed in 6 of 29 chronic alcoholic men given both drugs, but in none of those given placebo plus disulfiram (38). Metronidazole inhibits the ring oxidation of S (+) warfarin and significant bleeding can occur if the two drugs are taken together (39). Significant increase of hepatic clearance of metronidazole has been reported when the drug was taken together with phenobarbital (40, 41) or prednisone (41).

Dosage (42)

The dosage regimen of metronidazole is still far from optimal and better treatment regimens are acutely needed in rural areas.

Amoebiasis (acute amoebic dysentery and liver abscess)

Adults and children
30 mg/kg daily orally in three divided doses after meals for 8–10 days, or i.v. in three divided injections daily until the patient is able to take oral formulations.

Giardiasis

Adults
2 g once daily orally for 3 days.
Children
15 mg/kg daily orally in three divided doses for 5–10 days.

Dracontiasis

Adults and children
25 mg/kg daily orally for 10 days, with a daily maximum of 750 mg for children.

Preparations

Many preparations are available apart from those mentioned below.

Available as metronidazole

- Elyzol® (Dumex). Solution for infusion 5 mg/ml. Tablets 250 mg, 500 mg. Suppositories 500 mg, 1000 mg.
- Flagyl® (Rhône-Poulenc Rorer). Solution for infusion 5 mg/ml. Tablets 200 mg, 400 mg. Suppositories 500 mg, 1000 mg.
- Servizol® (Servipharm). Tablets 200 mg, 250 mg.

Available as metronidazole benzoate: 10 mg metronidazole benzoate is equivalent to 6.2 mg metronidazole.

- Elyzol (Dumex)® Oral solution 25 mg metronidazole base/ml.
- Flagyl® (Rhône-Poulenc Rorer). Oral solution 40 mg metronidazole base/ml.

References

1. Scully BE (1988). Metronidazole. *Med Clin North Amer*, **72**, 613–621.
2. Muller M (1983). Mode of action of metronidazole on anaerobic bacteria and protozoa. *Surgery*, **93**, 165–171.
3. Gulaid A, Houghton GW, Lewellen ORW, Smith J, Thorne PS (1978). Determination of metronidazole and its major metabolites in biological fluids by high pressure liquid chromatography. *Br J Clin Pharmacol*, **6**, 430–432.
4. Jensen JC, Gugler R (1983). Sensitive high-performance liquid chromatographic method for the determination of metronidazole and its metabolites. *J Chromatogr*, **277**, 381–384.
5. Ralph ED (1983). Clinical pharmacokinetics of metronidazole. *Clin Pharmacokinet*, **8**, 43–62.
6. Mattila J, Mannisto PT, Mantyla R, Nykanen S, Lamminsivu V (1983). Comparative pharmacokinetics of metronidazole and tinidazole as influenced by administration route. *Antimicrob Agents Chemother*, **23**, 721–725.
7. Lau AH, Emmons K, Seligsohn R (1991). Pharmacokinetics of intravenous metronidazole at different dosages in healthy subjects. *Int J Clin Pharmacol Ther Toxicol*, **29**, 386–90.
8. Houghton GW, Thorne PS, Smith J, Templeton R, Collier JA (1979). Comparison of the pharmacokinetics of metronidazole in healthy female volunteers following either a single oral or intravenous dose. *Br J Clin Pharmacol*, **8**, 337–341.
9. Lau AH, Lam NP, Piscitelli SS (1992). Clinical pharmacokinetics of metronidazole and other nitroimidazole anti-infectives. *Clin Pharmacokinet*, **23**, 328–364.
10. Stambaugh JE, Feo LG, Manthei RW (1968). The isolation and identification of the urinary oxidative metabolites of metronidazole in man. *J Pharmacol Exper Therapeut*, **161**, 373–381.
11. Loft S, Otton SV, Lennard MS, Tucker GT, Poulsen HE (1991). Characterization of metronidazole metabolism by human liver microsomes. *Biochem Pharmacol*, **41**, 1127–1134.
12. Jensen JC, Gugler R (1983). Single- and multiple-dose metronidazole kinetics. *Clin Pharmacol Ther*, **34**, 481–487.
13. Schwartz DE, Jeunet F (1976). Comparative pharmacokinetic studies of ornidazole and metronidazole in man. *Chemotherapy*, **22**, 19–29.
14. Bergan T, Thorsteinsson SB (1986). Pharmacokinetics of metronidazole and its metabolites in reduced renal failure. *Chemotherapy*, **32**, 305–318.
15. Houghton GW, Dennis MJ, Gabriel R (1985). Pharmacokinetics of metronidazole in patients with degrees of renal failure. *J Clin Pharmacol*, **19**, 203–209.
16. Loft S, Sonne J, Dossing M, Andreason PB (1987). Metronidazole pharmacokinetics in patients with hepatic encephalopathy. *Scan J Gastroenterol*, **22**, 117–123.
17. Somogyi A, Kong C, Sabto J, Gurr FW, Spicer WJ, McLean AJ (1983). Disposition and removal of metronidazole in patients undergoing haemodialysis. *Eur J Clin Pharmacol*, **25**, 683–687.
18. Amon I, Amon K, Scharp H, Franke G, Nagel F (1983). Disposition kinetics of metronidazole in children. *Eur J Clin*, **24**, 113–119.
19. Hall P, Kaye CM, McIntosh N, Steele J (1983). Intravenous metronidazole in the newborn. *Arch Dis Child*, **58**, 529–531.
20. Lares-Asseff I, Cravioto J, Santiago P, Perez-Ortiz B (1992). Pharmacokinetics of metro-nidazole in severely malnourished and nutritionally rehabilitated children. *Clin Pharmacol Ther*, **51**, 42–52.
21. Islam N, Hasan K (1978). Tinidazole and metronidazole in hepatic amoebiasis. *Drugs*, **15**, 26–29.
22. Bakshi JS, Ghiara JM, Nanivadekar AS (1978). How does tinidazole compare with metronidazole? A summary of report of Indian trials in amoebiasis and giardiasis. *Drugs*, **15**, 33–42.
23. Kokhani RC, Garud AD, Deodhar KP, Sureka SB, Kulkanni M, DamLe VB (1978). Treatment of amoebic liver abscess with tinidazole and metronidazole. *Drugs*, **15**, 23–25.
24. Simjee AE, Gathiram V, Jackson TFHG, Khan BFY (1985). A comparative trial of metronidazole v. tinidazole in the treatment of amoebic liver abscess. *S Afr Med J*, **68**, 923–924.
25. Bassilly S, Farid Z, El-Masry NA, Mikhail EM (1987). Treatment of intestinal *E. histolytica* and *Giardia lamblia* with metronidazole, tinidazole and ornidazole: a comparative study. *J Trop Med Hyg*, **90**, 9–12.
26. Gazder AJ, Banerjee M (1978). Single dose therapy of giardiasis with tinidazole and metronidazole. *Drugs*, **15**, 30–32.
27. Jokipii L, Jokipii ANM (1979). Single-dose metronidazole and tinidazole as therapy for giardiasis: success rates, side effects, and drug absorption and elimination. *J Infect Dis*, **140**, 984–988.

28. Padonu KO (1973). A controlled trial of metronidazole in the treatment of dracontiasis in Nigeria. *Am J Trop Med Hyg*, **22**, 42–44.
29. Kale OO, Elemile T, Enahoro F (1983). Controlled comparative trial of thiabendazole and metronidazole in the treatment of drancontiasis. *Am J Trop Med Parasitol*, **77**, 151–157.
30. Antani JA, Srinivas HV, Krishnamurthy KR, Borgaonakar AN (1972). Metronidazole in drancontiasis. *Am J Trop Med Hyg*, **21**, 178–181.
31. Roe FJC (1985). Safety of nitroimidazoles. *Scand J Infect Dis*, **46**, 72–81.
32. Heisterberg L (1984). Placental transfer of metronidazole and tinidazole in early human pregnancy after a single infusion. *Br J Clin Pharmacol*, **18**, 254–257.
33. Briggs GG, Freeman RK, Yaffe SJ (eds) (1994). *Drugs in Pregnancy and Lactation. A Reference Guide to Fetal and Neonatal Risk* (4th edn), (Baltimore: Williams & Wilkins), pp. 585/m–587/m.
34. Heisterberg L, Branebjerg PE (1983). Blood and milk concentrations of metronidazole in mothers and infants. *J Perinat Med*, **11**, 114–120.
35. White CM, Price JJ, Hunt KM (1980). Bone marrow aplasia associated with metronidazole. *BMJ*, **280**, 647.
36. Fagan TC, Johnson DG, Grosso DS (1985). Metronidazole-induced gynecomastia. *J Am Med Ass*, **254**, 3217.
37. Poltkin BH, Cohen I, Tsang T, Cullinane T (1985). Metronidazole-induced pancreatitis. *Ann Intern Med*, **103**, 891–892.
38. Rothstein E, Clancy DD (1969). Toxicity of disulfiram combined with metronidazole. *N Engl J Med*, **280**, 1006–1007.
39. O'Reilly RA (1976). The stereoselective interaction of warfarin and metronidazole in man. *N Engl J Med*, **295**, 354–357.
40. Gupte S (1983). Phenobarbital and metabolism of metronidazole. *N Engl J Med*, **308**, 529.
41. Eradiri D, Jamali R, Thomson ABR (1988). Interaction of metronidazole with phenobarbital, cimetidine, prednisone, and sulphasalzine in Crohn's disease. *Biopharmaceut Drug Disp*, **9**, 219–227.
42. *WHO Model Prescribing Information. Drugs used in parasitic diseases* (1990), (Geneva: World Health Organization).

Niclosamide

Chemical structure

Niclosamide

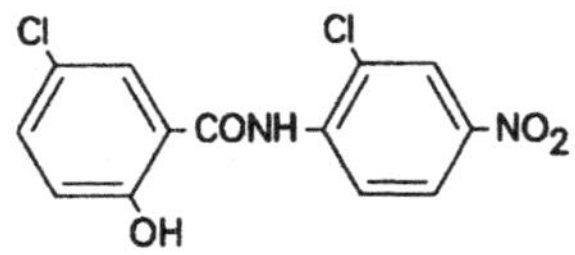

Physical properties

MW 327; pKa not known. Practically insoluble in water. Protect from light.

Pharmacology and mechanism of action

Niclosamide is a chlorinated salicylanilide derivative which was introduced during the 1960s. It is an anthelminthic drug highly effective against beef tapeworm (*Taenia (T.) saginata*), pork tapeworm (*T. solium*), fish tapeworm (*Diphyllobothrium (D.) latum*) and dwarf tapeworm (*Hymenolepis (H.) nana*).

The mechanism of action of the drug is not clearly known. It interferes with the energy metabolism of helminths, possibly by inhibiting adenosine triphosphate (ATP) production. It also inhibits glucose uptake by the parasites (1).

Pharmacokinetics

A specific analytical method has not been reported.

The pharmacokinetics of the drug is largely unknown and it seems that it is insignificantly absorbed.

Clinical trials

Most of the documentations on its efficacy are based on old and uncontrolled studies, however, niclosamide is generally regarded as a safe and effective drug.

After a single 2 g dose in patients infected with *T. saginata*, *T. solium* and *D. latum*, a cure rate of 90–100% has been reported (2, 3). Against *Hymenolepis nana*, a single dose of 70–130 mg/kg cured 100% of patients (4), although longer treatment periods were needed in some other studies (5, 6).

Indications

Infections caused by *T. saginata*, *D. latum*, and *H. nana*. The drug is also effective against *T. solium*, but the danger of cysticercosis makes praziquantel preferable.

Pregnancy and lactation

Niclosamide can be safely given to pregnant women if there is a strong indication for use, i.e. infections with *T. solium* because of the risk of cysticercosis.

Its excretion into breast milk is unknown, but it is unlikely to have of any clinical significance.

Side effects

Niclosamide is generally free of undesirable effects. Minor gastrointestinal complaints such as nausea, vomiting, abdominal pain, diarrhoea, light-headedness and pruritus are rarely encountered.

Contraindications and precautions

Drugs which cause vomiting should not be taken simultaneously with niclosamide to avoid retrograde infection by eggs. Use of niclosamide against *T. solium* infection may expose the patient to the risk of cysticercosis. In such a case, special precautions are needed (see Dosage below).

Interactions

There have been no reports.

Dosage

D. Latum, T. saginata, T. solium

Adults
2 g in the morning.
Children 11–34 kg
1 g in the morning.
Children >34 kg
1.5 g in the morning.

In order to avoid the danger of cysticercosis in the treatment of *T. solium*, an antiemetic before and a purge should be given within 3 to 4 hours of the drug being given.

H. nana

Adults
A single daily dose of 2 g for 7 days.
Children 11–34 kg
A single 1 g dose on day 1. Thereafter a single daily dose of 0.5 g for 6 days.
Children >34 kg
A single 1.5 g dose on day 1. Thereafter a single daily dose of 1 g for 6 days.

The tablets should be chewed and swallowed with some water after breakfast.

Preparations

- Niclocide® (Miles). Tablets 500 mg.
- Trédémine® (Bellon). Tablets 500 mg.
- Yomesan® (Bayer). Tablets 500 mg.

References

1. Webster LT Jr (1990). Drugs used in the chemotherapy of helminthiasis. In: *Goodman & Gilman's The Pharmacological Basis of Therapeutics*, 8th edn, edited by A. G. Gilman, T. W. Rall, A. S. Nies, P. Taylor, (New York: Pergamon Press), pp. 965–966.
2. Perera DR, Western KA, Schultz MG (1970). Niclosamide treatment of cestodiasis. Clinical trial in the United States. *Am J Trop Med Hyg*, **19**, 610–612.
3. Schneider J (1963). Treatment of *Taenia saginata* infection with niclosamide-5-chloro-N-(2-chloro-4-nitrophenyl) salicylamide. *Bull Soc Path Exot*, **56**, 451–461.
4. Ahkami S, Haijan A (1970). Radical treatment of *Hymenolepis nana* with niclosamide. *J Trop Med Hyg*, **73**, 258–259.
5. El-Masry NA, Farid Z, Bassily S (1974). Treatment of *Hymenolepis nana* with niclosamide, mepacrine, and thiabendazole. *East Afr Med*, **51**, 532–535.
6. Most H, Yoeli M, Hammond J, Scheinesson GP (1971). Yomesan (niclosamide) therapy of *Hymenolepis nana* infections. *Am J Trop Med Hyg*, **20**, 206–208.

Nifurtimox

Chemical structure

Nifurtimox

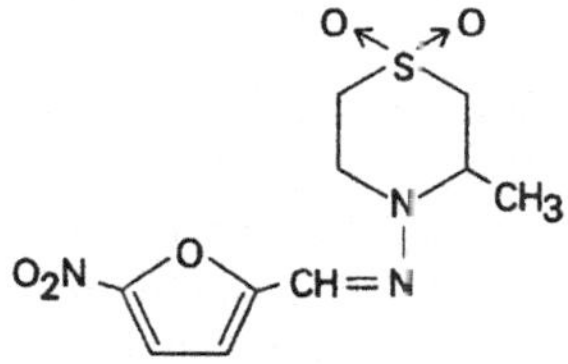

Physical properties

MW 287; pKa not known. Slightly soluble in water. Protect from light. The drug should be stored in air-tight containers.

Pharmacology and mechanism of action

Nifurtimox is a nitrofuran derivative that has trypanocidal activity against both the trypomastigote forms (extracellular) and the amastigote forms (intracellular) of *Trypanosoma (T.) cruzi*. Under experimental conditions amastigotes are 10 times more sensitive to the drug than the trypomastigotes (1).

The mechanism of action of the drug is not clearly known. Its trypanocidal action may be related to its ability to undergo partial reduction to form chemically reactive radicals that cause production of superoxide anion, hydrogen peroxide and hydroxyl radicals. These free radicals react with cellular macromolecules and cause membrane injury, enzyme inactivation, damage to DNA, and mutagenesis (2).

Pharmacokinetics

A specific HPLC method has been described for the determination of nifurtimox (3).

Nifurtimox is given orally. Its bioavailability in man is unknown, but based on animal studies the drug is likely to be completely absorbed (4). In healthy human volunteers given single oral doses of 15 mg/kg of the drug, average peak plasma levels of 751 ng/ml (range 356–1093 ng/ml) were reached within 2–3 hours. The drug was distributed with an apparent volume of distribution of about 755 l and was quickly eliminated with an average plasma elimination half-life of 3 hours (range 2–6 hours) (5). Nifurtimox has been reported to be extensively metabolized in animals including man, but the nature of its metabolic products is not known. For all dosages studied in man, dogs, and rats, less than 1% of the orally administered dose was excreted with the urine as the parent drug (4). Higher concentrations of the drug were reported in patients with kidney failure compared to normal healthy volunteers but these patients may also have had concomitant liver diseases (6).

Clinical trials

Treatment of acute stages of *T. cruzi* infections results in the disappearance of parasitaemia and amelioration of symptoms in about 80% of patients (7). In the treatment of chronic forms over 90% cure rates have been reported from open trials performed in Argentina, southern Brazil, Chile and Venezuela (8). Less satisfactory results were found in studies carried out in central Brazil (8). Diagnostic techniques of trypanosomiasis such as xenodiagnosis and serological tests have low sensitivity which may have affected the outcome of the trial. Compliance and rate of re-infection are also common problems in such trials, especially in ambulatory patients.

Studies on the efficacy of nifurtimox against African trypanosomiasis is limited to a few studies on late stage gambiense sleeping sickness and the results are inconsistent. In one study in Zaire (9), the drug was reported to have cured 7 of 15 patients treated with dosages of 4–5 mg/kg 3 times a day during 2 months. Children received 20 mg/kg per day. Follow-up was 30 months. In a similar study carried out in another area in Zaire (10), the drug was reported to have cured none of 20 patients who were followed-up from 1 to 9 months. In two studies conducted in the Sudan (11), the drug was reported to have cured 60 of 95 patients (63%) treated with 5 mg/kg 3 times a day for 14–45 days. Children received 20 mg/kg. Follow-up was 4 months. Side effects were common, but most patients were in bad condition prior to nifurtimox. The studies were open trials and there was little control of drug intake. The discrepancies between the studies may be due to poor compliance.

Several open clinical trials have also shown that nifurtimox is effective against some cases of cutaneous and mucocutaneous leishmaniasis (12, 13). However, most patients showed side effects, and the drug can not be recommended for routine use in either type of leishmaniasis (14).

Indications

Treatment of American trypanosomiasis (Chagas' disease) due to *Trypanosoma cruzi*. The drug may also be used in patients with *Trypanosoma brucei gambiense* sleeping sickness who are refractory to other treatments.

Pregnancy and lactation

Teratogenicity has been reported in rats and mice (15). Documentation in man is lacking. The drug should not be withheld from pregnant women with acute *Trypanosoma cruzi* infection. In chronic cases, treatment may be postponed until after the first trimester.

Its excretion into breast milk is unknown.

Side effects

Side effects of nifurtimox are frequent and can be encountered in up to 40% in children, and up to 70% in adults treated for acute and chronic Chagas' disease. Common side effects include anorexia, nausea, vomiting, abdominal pain, excitation, sleeping difficulties, dizziness, headache and joint and muscle pains (16). During treatment, half of the patients may interrupt therapy because of side effects. Other rare side effects include skin eruptions and paraesthesia (7).

Contraindications and precautions

The drug should be given with caution to patients with a history of convulsions, brain injury, peripheral neuropathy and psychiatric illness. Dosage reductions may be considered in patients with liver diseases.

Interactions

Concomitant administration of nifurtimox with melarsoprol (17) or eflornithine (18) have been reported to have synergistic effects in experimental animals (mice) infected with *Trypanosoma brucei* species. The clinical implication of this is unknown.

Dosage (19)

The treatment period of nifurtimox is long and is largely based on clinical experience. With such a treatment schedule and the fact that the drug is toxic, it is unlikely that patients will complete the treatment. The short half-life of the drug necessitating frequent intake also complicates the drug regimen. A slow release preparation may have been suitable in this case.

Adults

8–10 mg/kg orally in 3 divided daily doses for 90 days.

Children

15–20 mg/kg orally in 4 divided daily doses for 90 days.

Preparations

- Lampit® (Bayer). Tablets 30 mg, 120 mg.

References

1. Webster LT Jr (1990). Drugs used in the chemotherapy of profozoal infections. In: *Goodman & Gilman's The Pharmacological Basis of Therapeutics*, 8th edn, edited by A. G. Gilman, T. W. Rall, A. S. Nies, P. Taylor, (New York: Pergamon Press), pp. 1010–1011.
2. Docampo R, Moreno SNJ, Stoppani AOM, Leon W, Cruz FS, Villalta F, Muniz RFA (1981). Mechanism of nifurtimox toxicity in different forms of *Trypanosoma cruzi*. *Biochem Pharmacol*, **30**, 1947–1981.
3. Paulos C, Paredes J, Vasquez I, Kunze G, Gonzalez-Martin G (1988). High performance liquid chromatographic determination of nifurtimox in human serum. *J Chromatogr*, **433**, 359–362.
4. Medenwald H, Brandau K, Schlossmann K (1972). Quantitative determination of nifurtimox in body fluids of rat, dog and man. *Arzneimittelforschung*, **22**, 1613–1616.
5. Paulos C, Paredes J, Vasquez I, Thambo S, Arancibia A, Gonzalez-Martin G (1989). Pharmacokinetics of a nitrofuran compound, nifurtimox, in healthy volunteers. *Int J Clin Pharmacol Ther Toxicol*, **27**, 454–457.
6. Gonzalez-Martin G, Thambo S, Paulos C, Vasquez I, Paredes. J (1992). The pharmacokinetics of nifurtimox in chronic renal failure. *Eur J Clin Pharmacol*, **42**, 671–674.
7. Wegner DHG, Rohwedder RW (1972). The effects of nifurtimox in acute Chagas' infection. *Arzneimittelforschung*, **22**, 1624–1635.
8. Wegner DHG, Rohwedder RW (1972). Experience with nifurtimox in chronic Chagas' infection. *Arzneimittelforschung*, **22**, 1635–1642.
9. Moens F, De Wilde M, Ngato K (1984). Essai de traitement du nifurtimox de la Trypanosomiase humaine Africaine. *Ann Soc Belge Med Trop*, **64**, 37–43.
10. Pepin J, Milord F, Mpia B, Meurice F, Ethier L, DeGroof D, Bruneel H (1989). An open clinical trial of nifurtimox for arseno-resistant *Trypanosoma brucei gambiense* sleeping sickness in central Zaire. *Trans R Soc Trop Med Hyg*, **83**, 514–517.

11. Van Nieuwenhove S (1992). Advances in sleeping sickness therapy. *Ann Soc Belg Med Trop*, **72**, 39–51.
12. Guerra MFV, Marsden PD, Cuba CC, Barretto AC (1981). Further trials of nifurtimox in mucocutaneous leishmaniasis. *Trans R Soc Trop Med Hyg*, **75**, 335–337.
13. Marsden PD, Cuba CC, Barretto AC, Sampaio RN, Rocha RA (1979). Nifurtimox in the treatment of South American leishmaniasis. *Trans R Soc Trop Med Hyg*, **73**, 391–394.
14. *Control of leishmaniasis*. WHO Tech Report Series no. 793 (1990). (Geneva: World Health Organization).
15. Lorke D (1972). Embryotoxicity studies of nifurtimox in rats and mice and study of fertility and general reproductive performance. *Arzneimittelforschung*, **22**, 1603–1612.
16. Gutteridge WE (1985). Existing chemotherapy and its limitations. *Br Med Bull*, **41**, 162–168.
17. Jennings FW (1991). Chemotherapy of CNS-trypanosomiasis: the combined use of the arsenicals and nitro-compounds. *Trop Med Parasitol*, **42**, 139–142.
18. Jennings FW (1988). The potentiation of arsenicals with difluoromethylornithine (DFMO): experimental studies in murine trypanosomiasis. *Bull Soc Pathol Exot*, **81**, 595–607.
19. *WHO Model Prescribing Information. Drugs used in parasitic diseases* (1990), (Geneva: World Health Organization).

Oxamniquine

Chemical structure

Oxamniquine

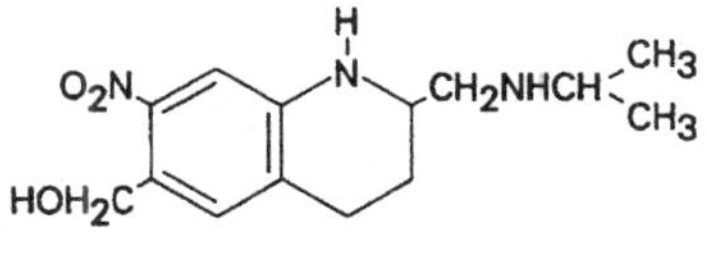

Physical properties

MW 279; pKa not known. The drug is almost insoluble in water. The capsules should be stored in air-tight containers.

Pharmacology and mechanism of action

Oxamniquine is a tetrahydroquinoline derivative effective in the treatment of *Schistosoma(s) mansoni* infections. The male worms are more susceptible to the drug effects than the female ones. It has no therapeutic value against other *Schistosoma* species. In experimental animal models, the drug causes a shift of the worms from the mesenteric veins to the liver where the male and the female decouple. The male worms preferentially concentrate the drug and die in the liver. The unpaired females return to the mesenteric vessels where they cease laying eggs and eventually die (1).

The mechanism of action of oxamniquine is unknown. The drug may induce its action by inhibiting DNA synthesis. When the drug was administered to rats infected with *S. mansoni*, it inhibited the synthesis of macromolecules in sensitive parasites and not in the resistant ones (2).

Pharmacokinetics

Specific GC (3) and HPLC (4) analytical methods have been described.

The drug is given orally. It is apparently well absorbed from the gastrointestinal tract (3). Peak plasma levels are reached between 1 and 4 hours after drug intake (4, 5). The drug is extensively metabolized in the body by oxidation to inactive metabolites. In healthy human volunteers given 600 mg of oxamniquine, 0.4–1.9% of the parent drug and 41–73% of a 6-carboxy metabolite (formed by oxidation product of the 6-hydroxymethyl group) were recovered in the urine over 36 hours. A small amount of a 2-carboxylic acid derivative (oxidation of the side chain) was also excreted during the same period (3). The drug is eliminated with a half-life of around 2 hours (4, 5). No significant differences were found in the pharmacokinetic parameters of oxamniquine when the drug was given to a small number of healthy volunteers and patients with advanced hepatosplenic schistosomiasis (4).

Clinical trials

Early clinical trials have revealed oxamniquine to be highly effective (100% cure rate) in curing acute as well as chronic *S. mansoni* infections. Doses used in those early studies were 7.5 mg/kg given intramuscularly. However, this route of administration has been abandoned because of moderate to severe local pain at the site of injection which persisted for more than a week (6–8). Later trials with oral oxamniquine have shown the oral route to be as effective as the parenteral route. However, differences on the efficacy of oxamniquine in different countries and regions have appeared. This has been reported to be due to differences in the sensitivity of the parasites in various regions to the drug (9–12).

It has also been reported that children need higher doses than adults and that they seem to be more tolerant to the adverse effects of the drug than adults (9, 10, 12). This might be due to differences in the absorption or metabolism of oxamniquine between adults and children.

In some community-based treatment programmes oxamniquine was reported to reduce the prevalence of *S. mansoni* from 63% to only 17% during an 8-year period and that of palpable livers and spleens from 87% to 31% and from 20% to 3%, respectively (13). A number of community-based chemotherapy programmes have been reviewed recently by Foster (14).

Indications

Oxamniquine is used against *S. mansoni* infections, including advanced cases with hepatomegaly, ascites or with colonic polyposis.

Pregnancy and lactation

Teratogenicity has not been reported in rats and rabbits (15). Documentation in man is lacking. Treatment with oxamniquine should be postponed until delivery, unless there is a strong indication for its use.

Its excretion into breast milk is unknown.

Side effects

Oxamniquine is generally well tolerated even during large scale treatment programmes. The only significant common side effect reported is mild to moderate dizziness with or without drowsiness, reported by up to 40% of treated patients. It starts up to 3 hours after a dose and usually lasts for 3 to 6 hours. Other side effects include nausea, vomiting, abdominal pain, and diarrhoea (14). Transient fever, 38 to 39 °C, peripheral blood eosinophilia and pulmonary infiltrates (Loeffler's syndrome) have been reported mainly from Egyptian patients 24 to 72 hours after completing a 3-day course of therapy (16). The cause seems to be unknown. A number of reports of epileptiform convulsions have been reported in patients suspected with (17) or without (18, 19) a history of epilepsy. More severe neuropsychiatric symptoms such as severe headache, hallucinations, episodes of fainting, severe amnesia, total disorientation in space and time and confusion have been rarely reported (12, 20).

Discoloration of the urine from orange to red may follow after the drug treatment (most likely due to a metabolite) (11). This is transitory and harmless, nevertheless patients should be informed about it.

Contraindications and precautions

Patients with pre-existing central nervous system disturbances such as epilepsy or psychiatric disorders should be treated with caution.

Dosage (21).

West Africa, South America, and the Caribbean islands

Adults
A single dose of 15 mg/kg.
Children (<4 years)
A single total dose of 20 mg/kg or two doses of 10 mg/kg in one day separated by an interval of 3 to 8 hours.

East and Central Africa (Kenya, Madagascar, Malawi, Rwanda, Burundi, Tanzania, Zambia) and the Arabian peninsula

Adults and children
30 mg/kg, given as 15 mg/kg twice daily for one day or once daily for two consecutive days.

Sudan, Uganda and Zaire

Adults and children
A total dose of 40 mg/kg.

Egypt, South Africa, and Mozambique

Adults and children
A total dose of 60 mg/kg, given as 15 mg/kg twice daily for two days, or 20 mg/kg once daily for 3 consecutive days.

Preparations

- Vansil® (Pfizer). Capsules 250 mg. Oral suspension 50 mg/ml.
- Mansil® (Pfizer). Capsules 250 mg. Oral suspension 50 mg/ml.

References

1. Webster LT Jr (1990). Drugs used in the chemotherapy of helminthiasis. In: *Goodman & Gilman's The Pharmacological Basis of Therapeutics*, 8th edn, edited by A. G. Gilman, T. W. Rall, A. S. Nies, P Taylor, (New York: Pergamon Press), pp. 966.
2. Pica-Mattoccia L, Coli D (1985). Studies on the mode of action of oxamniquine and related schistosomicidal drugs. *Am J Trop Med Hyg*, **34**, 112–118.
3. Kaye B, Woolhouse NM (1976). The metabolism of oxamniquine, a new schistosomicide. *Ann Trop Med Parasitol*, **70**, 323–328.
4. Daneshmend TK, Homeida M (1987). Oxamniquine pharmacokinetics in hepatosplenic schistosomiasis in the Sudan. *J Antimicrob Chemother*, **19**, 87–93.
5. Kokwaro GO, Taylor G (1991). Oxamniquine pharmacokinetics in healthy Kenyan African volunteers. *East Afr Med J*, **68**, 359–364.
6. Rodrigues C, Argento CA, de Figueiredo N, Wanke B, de Queiroz GC (1973). Clinical trial with oxamniquine-(U.K. 4271)-in the treatment of schistosomiasis mansoni. *Rev Inst Med Trop São Paulo*, **15**, 41–46.

7. Coutinho A, Domingues AC, Bonfim RA (1973). Treatment of *mansoni* schistosomiasis with oxamniquine. *Rev Inst Med Trop São Paulo*, **15**, 15–34.
8. Katz N, Pellegrino J, Grinbaum E, Chaves A, Zicker F (1973). Further clinical trials with oxamniquine, a new antischistosomal agent. *Rev Inst Med Trop São Paulo*, **15**, 35–40.
9. Katz N, Zicker F, Pereira JP (1977). Field trials with oxamniquine in a schistosomiasis mansoni-endemic area. *Am J Trop Med Hyg*, **26**, 234–237.
10. Kilpatrick ME, Farid Z, Bassily S, El-Masry NA, Trabolsi B, Watten RH (1981). Treatment of *Schistosoma mansoni* with oxamniquine-five years' experience. *Am Soc Trop Med Hyg*, **30**, 1219–1222.
11. Omer AHS (1978). Oxamniquine for treating *Schistosoma mansoni* infection in Sudan. *BMJ*, **2**, 163–165.
12. Katz N, Grinbaum E, Chaves A, Zicker F, Pellegrind J (1976). Clinical trials with oxamniquine by oral route, in *Schistosoma mansoni*. *Rev Inst Med Trop São Paulo*, **18**, 371–377.
13. Sleigh AC, Hoff R, Mott KE, Maguire JH, da França-Silva JT (1986). Manson's schistosomiasis in Brazil: 11-year evaluation of successful disease control with oxamniquine. *Lancet*, **i**, 635–637.
14. Foster R (1987). A review of clinical experience with oxamniquine. *Trans R Soc Trop Med Hyg*, **81**, 55–59.
15. Oxamniquine. *Therapeutic Drugs*, edited by Sir Colin Dollery (1990), (London: Churchill Livingstone), pp: O42-O45.
16. Higashi GI, Farid Z (1979). Oxamniquine fever: drug-induced or immune-complex reaction? *BMJ*, **ii**, 830.
17. Krajden S, Keystone JS, Glenn C (1983). Safety and toxicity of oxamniquine in the treatment of *Schistosoma mansoni* infections, with particular reference to electroencephalographic abnormalities. *Amer J Trop Med Hyg*, **32**, 1344–1346.
18. Stockvis H, Bauer AGC, Stuiver PC, Malcolm AD, Overbosche D (1986). Seizures associated with oxamniquine therapy. *Am J Trop Med Hyg*, **35**, 330–331.
19. Al-aska AK (1985). Treatment of *Schistosoma mansoni* infection with oxamniquine in Riyadh, Saudi Arabia. *Trop Med Parasitol*, **36**, 213–214.
20. Chunge CN, Kimani RG, Gachihi G, Mkoji G, Kamau T, Rashid JR (1985). Serious side effects of oxamniquine during the treatment of *Schistosoma mansoni* in Kenya. *East Afr Med J*, **62**, 3–4.
21. Anthelminthics. *Martindale: The Extra Pharmacopoeia*, 30th edn (1993), (London: Pharmaceutical Press), pp: 4950.

Pentamidine

Chemical structure

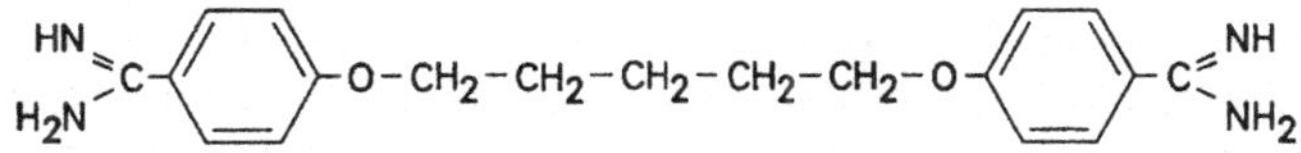

Physical properties

Base: MW: 341; isethionate; MW 593; pKa: 11.4. 1 g of the isethionate dissolves in 10 ml of water. Parenteral solutions of pentamidine deteriorate on storage and should be used within few hours after preparation.

Pharmacology and mechanism of action

Pentamidine is a synthetic aromatic diamidine chemically related to the antidiabetic drug phenformin and was introduced into the treatment of trypanosomiasis and leishmaniasis in 1940. Pentamidine has also an established place in the treatment of *Pneumocystis carinii* pneumonia (1).

The mechanism of action of pentamidine is not known. The drug concentrates in trypanosomes via an energy-dependent, high affinity uptake system, which operates more rapidly in drug sensitive strains. Inside the cell, pentamidine interacts with nucleic acids thus affecting DNA biosynthesis (2). It has also been shown that the drug inhibits the plasma-membrane Ca^{++}-ATPase of the parasites (3). *In vitro*, the drug causes ultra-structural disruptions of the mitochondrial structures of *Leishmania mexicana* and *tropica* (2, 4). The drug has also been shown to inhibit trypanosomal S-adenosyl-L-methionine decarboxylase, thus reducing the synthesis of polyamines (5).

Pharmacokinetics

Specific HPLC methods have been described for the determination of pentamidine (6–9).

The drug is given parenterally because of its low oral bioavailability. Using specific analytical methods, the pharmacokinetics of pentamidine have only recently been studied in patients with AIDS (10) or with trypanosomiasis (11). These initial studies suffered from short sampling periods of 24 to 48 hours which does not allow proper pharmacokinetic evaluations. Recently, a more detailed single dose pharmacokinetic study of pentamidine in 11 patients with *Trypanosoma brucei gambiense* has been reported (12). The patients were administered a 2 hour intravenous infusion of 3.0 to 4.8 mg/kg of pentamidine isethionate. Plasma concentrations of pentamidine were measured up to 1–8 months after dosing. Maximal plasma concentrations were attained at the end of the infusion and varied three-fold. After termination of infusion, the drug disappeared with a multiple exponential elimination pattern. The average terminal elimination half-life was 265 (range 107–446) hours. The renal clearance accounted for only about 11% of the total plasma clearance, suggesting metabolic clearance of the drug (12). In a multiple dose study (13), pentamidine isethionate was studied in 12 patients with *Trypanosoma brucei gambiense* given 2-hour intravenous infusions every other day for 20 days. The mean dose per infusion varied between 3.8 and 4.2 mg/kg of pentamidine isethionate. The trough plasma concentrations of the drug increased gradually indicating drug accumulation. The median volume of distribution at steady state was 8500 l.

Pentamidine binds strongly and extensively to lysosomes (14) which is consistent with the reported large volume of distribution (12). About 70% of the drug is bound to plasma proteins (Ericsson *et al.* unpublished work). Less than 1% of the concentrations of the drug in plasma were measured in CSF (11), but the clinical relevance of this is unknown. Pentamidine has been shown to be metabolized in rats (15, 16), and in human liver microsomes (17).

Clinical trials

Trypanosoma brucei gambiense

The first clinical trials of pentamidine against trypanosomiasis were conducted more than 50 years ago. In the first recorded large study, a cure rate of 90 to 100% was reported in individuals without CNS involvement. The patients were generally treated with 0.8–1.7 mg/kg pentamidine hydrochloride given intravenously once daily for 8–12 days. The patients were followed for more than one year after treatment (18). Similar results were reported by later studies (19). Cure rates of less than 40% were reported in patients described to have CNS involvement prior to treatment. Jonchère reported a cure rate of 93% in 9100 patients who were followed up for 4 years. The patients were treated with 5 injections of 2–6 mg/kg pentamidine base intramuscularly once every other day. The majority of the patients who relapsed (7%) had CNS involvement before treatment (20).

Pentamidine has also some effect against *Trypanosoma brucei rhodesiense* without CNS involvement, but it is considered to be less effective than suramin (21).

Leishmaniasis

In an open clinical study, pentamidine isethionate (4 mg/kg on alternate days) was administered i.m. to patients with visceral leishmaniasis (kala-azar) (resistant to sodium antimony gluconate). All but one of the 82 patients responded to the therapy (22). In a similar study in India, 240 patients with visceral leishmaniasis unresponsive to antimonials were treated with pentamidines 4 mg/kg on alternate days for 20 days. 175 of them received pentamidine isethionate, while 65 received pentamidine dimesylate. One month after treatment, clinical and parasitological cures were 75% and 100% respectively. However, the dimesylate salt was observed to be more toxic (23). In another study in India, 312 patients with antimony-resistant kala-azar were randomized into 3 treatment groups: pentamidine isethionate (4 mg/kg i.v. 3 times per week until 11 weeks); similar dose of pentamidine isethionate given concomitantly with a 20-day regimen of sodium stibogluconate; and similar dose of pentamidine isethionate followed by 20 days of sodium stibogluconate therapy. Six months after treatment, cure rates observed were 78, 84 and 98%, respectively (24). In Colombia, 92 patients with cutaneous leishmaniasis were randomly treated either with meglumine antimonate (10 mg Sb/kg i.m. twice a day for 20 days), pentamidine isethionate (2 mg/kg every other day i.m. for 14 days), itraconazole (200 mg orally, twice a day for 28 days) or no treatment. Fifteen months after treatment, the group given pentamidine had a cure rate of 96%, which was superior to that of the other groups (25). In a prospective randomized trial, pentamidine isethionate (4 mg/kg i.m. on alternate days for 40 days) was compared with amphotericin (0.5 mg/kg i.v. on alternate days for 28 days) in 120 patients with uncomplicated cases of antimony-unresponsive kala-azar. Six months after treatment pentamidine was less effective than amphotericin with cure rates of 77% and 98%, respectively (26).

Indications

Trypanosomiasis

In the treatment of early-stage *Trypanosoma brucei gambiense*.

Leishmaniasis

In the treatment of patients with visceral, diffuse cutaneous, or mucocutaneous leishmaniasis due to *L. aethiopica* and *L. guyanensis* who are unresponsive or intolerant to antimony preparations.

Pneumocystis carinii pneumonia

Pentamidine is also used in the prophylaxis and treatment of *Pneumocystis carinii* pneumonia as a second choice.

Pregnancy and lactation

Teratogenicity has not been reported in rabbits (27). Documentation in man is lacking, despite the wide spread use of the drug for many years. Because of the severity of visceral leishmaniasis and trypanosomiasis, pentamidine should be given to pregnant women suffering from such diseases.

Its excretion into breast milk is unknown.

Side effects

Following parenteral administration of pentamidine, about 45% of the patients may experience side effects some of which can be fatal (1, 2). Rapid i.v. injection may cause sudden hypotension followed by breathlessness, tachycardia, dizziness, headache, vomiting and fainting which are due to histamine release. Local pain and sterile abscess may be formed after intramuscular injection (28). Nephrotoxicity which is usually mild to moderate and reversible is the most common side effect. Hypoglycaemic reactions are also common. A few patients may develop hyperglycaemia and diabetes mellitus (1, 2). This paradoxical effect is thought to be due to a cytolytic release of insulin followed by destruction of the beta-cells. Leucopenia, abnormal liver function, hypocalcaemia and Stevens–Johnson syndrome can also occur (1, 2). There have been occasional reports of acute pancreatitis (29).

In two smaller studies conducted in Côte d'Ivoire, the most common subjective side effect reported was abdominal pain. A few patients complained of hypersalivation (12). Pentamidine has been reported to possess anticholinesterase activity (30) and the abdominal pain and the hypersalivation reported in the former study may be due to this effect.

Contraindications and precautions

Patients should remain supine during pentamidine administration and their blood pressure and blood glucose levels monitored. It is advisable to have adrenaline ready in case of a sudden collapse. Where possible, kidney and liver function tests should be assessed regularly during the treatment.

Pentamidine is not recommended for use in patients with late-stage trypanosomiasis.

Interactions

There have been no reports.

Dosage and administration (31)

The presently used dosage regimen of pentamidine is complicated and is solely based on clinical experience. In the light of the new pharmacokinetic findings of the drug, a more simplified regimen is likely to be designed in the future.

Pentamidine might be administered intramuscularly (i.m.) or intravenously (i.v.). When given i.v., the drug should be infused for at least 60 minutes and the patient placed in a supine position. Adrenaline should also be at hand in case of acute reaction. Adults and children are given the same dose per kg. The dosage of pentamidine is calculated as the salt (isethionate).

Trypanosoma brucei gambiense

4 mg/kg i.m. daily for 7–10 days. It is essential that the cerebral type of the disease is excluded.

Visceral leishmaniasis

4 mg/kg i.m. 3 times weekly for 5–25 weeks or longer depending on the clinical and parasitological cure of the patient.

Cutaneous leishmaniasis (caused by L. aethiopica and L. guyanensis)

3–4 mg/kg i.m. once or twice weekly until the lesion is no longer visible. Relapse is unusual.

Diffuse cutaneous leishmaniasis (L. aethiopica)

3–4 mg/kg i.m. once a week, continuously for at least 4 months until parasites are no longer detectable in slit-skin smears. Relapse frequently occurs during the first few months before immunity is established.

Mucocutaneous leishmaniasis (L. braziliensis and L. aethiopica)

4 mg/kg i.m. three times a week for 5–25 weeks, or until the lesion is no longer visible.

Preparations

Available as pentamidine isethionate (174 mg isethionate is equivalent to 100 mg base).

- Pentacarinat® (Rhône-Poulenc Rorer). Vials containing 300 mg pentamidine isethionate powder which is dissolved with 2–3 ml of sterile water.

- Pentam-300® (LyphoMed, USA). Vials containing 300 mg pentamidine isethionate powder which is dissolved with 2–3 ml of sterile water.

References

1. Goa KL, Campoli-Richards D (1987). Pentamidine Isethionate. A review of its antiprotozoal activity, pharmacokinetic properties and therapeutic use in *Pneumocystis carinii* pneumonia. *Drugs*, **33**, 242–258.
2. Sands M, Kron MA, Brown RB (1985). Pentamidine: A review. *Rev Infect Dis*, **7**, 625–634.
3. Benaim G, Lopez-Estrano C, Docampo R, Moreno SNJ (1993). A calmodulin-stimulated Ca^{2+} pump in plasma-membrane vesicles from *Trypanosoma brucei*; selective inhibition by pentamidine. *Biochem J*, **296**, 756–763.
4. Steck EA, Kinnamon KE, Rane DS, Hanson WL (1981). *Leishmania donovani, Plasmodium berghei, Trypanosoma rhodesiense*: antiprotozoal effects of some amidine types. *Exp Parasitol*, **52**, 404–413.
5. Bitonti AJ, Dumont JA, McCann PP (1986). Characterisation of *Trypanosoma brucei brucei* S-adenosyl-L-methionine decarboxylase and its inhibition by Berenil, pentamidine and methylglyoxal bis(quanylhydrazone). *Biochem J*, **237**, 518–521.
6. Lin JM, Shi RJ, Lin ET (1986). High performance liquid chromatographic determination of pentamidine in plasma. *J Liq Chromatogr*, **9**, 2035–2046.
7. Dusci LJ, Hackett LP, Forbes AM, Ilett KF (1987). High performance liquid chromatographic method for measurement of pentamidine in plasma and its application in an immunosuppressed patient with renal dysfunction. *Ther Drug Monit*, **9**, 422–425.
8. Ericsson Ö, Rais M (1990). Determination of pentamidine in whole blood, plasma and urine by high-performance liquid chromatography. *Ther Drug Monit*, **12**, 362–365.
9. Yeh TK, Dalton JT, Au JL (1993). High-performance liquid chromatographic determination of pentamidine in plasma. *J Chromatogr*, **622**, 255–261.
10. Conte JE Jr, Upton RA, Phelps RT, Wofsy CB, Zurlinden E, Lin ET (1986). Use of a specific and sensitive assay to determine pentamidine pharmacokinetics in patients with AIDS. *J Infect Dis*, **154**, 923–929.
11. Bronner U, Doua F, Ericsson Ö, Gustafsson LL, Miezan TW, Rais M, Rombo L (1991). Pentamidine concentrations in plasma, whole blood and cerebrospinal fluid during treatment of *Trypanosoma gambiense* infection in Côte d'Ivoire. *Trans R Soc Trop Med Hyg*, **85**, 608–611.
12. Bronner U, Gustafsson LL, Doua F, Ericsson Ö, Miézan T, Rais M, Rombo L (1995). Pharmacokinetics and adverse reactions of a single dose of pentamidine in patients with *Trypanosoma gambiense* sleeping sickness. *Br J Clin Pharmacol*, **39**, 289–295.
13. Bronner U, Rombo L, Doua F, Ericsson Ö, Miézan T, Gustafsson LL. Multiple-dose pharmacokinetics and adverse reactions of pentamidine in patients with *Trypanosoma gambiense* sleeping sickness. (In manuscript).
14. Glaumann H, Bronner U, Ericsson Ö, Gustafsson LL, Rombo L (1994). Pentamidine accumulates in rat liver lysosomes and inhibits phospholipid degradation. *Pharmacol Toxicol*, **74**, 17–22.
15. Berger BJ, Reddy VV, Le ST, Lombardy RJ, Hall JE, Tidwell RR (1991). Hydroxylation of pentamidine by rat liver microsomes. *J Pharmacol Exp Ther*, **256**, 883–889.
16. Bronner U, Ericsson Ö, Nordin J, Wikström I, Aden Abdi Y, Hall JE, Tidwell RR, Gustafsson LL (1995). Metabolism is an important route of pentamidine elimination in the rat: disposition of ^{14}C-pentamidine and identification of metabolites in urine using liquid chromatography-tandem mass spectrometry. *Pharmacol Toxicol*, **77**, 114–120.
17. Clement B, Jung F (1994). N-hydroxylation of the antiprotozoal drug pentamidine catalyzed by rabbit liver cytochrome P-450 2C3 or human liver microsomes, microsomal retroreduction, and further oxidative transformation of the formed amidoximes. Possible relationship to the biological oxidation of arginine to NG-hydroxyarginine, citrulline, and nitric oxide. *Drug Metab Dispos*, **22**, 486–497.
18. Lourie EM (1942). Treatment of sleeping sickness in Sierra Leone. *Ann Trop Med Parasitol*, **36**, 113–131.
19. Harding RD (1944). Trypanosomiasis treated with with pentamidine. *BMJ*, **2**, 447.
20. Jonchère H (1958). Traitement par les diamidines de la phase lymphatico-sanguine de la trypanosomiase humaine en A.O.F. *Bull Soc Path Exot Fil*, **44**, 603–625.

21. Apted FIC (1980). Present status of chemotherapy and chemoprophylaxis of human trypanosomiasis in the eastern hemisphere. *Pharmacol Ther*, **11**, 391–413.
22. Jha TK (1983). Evaluation of diamidine compound (pentamidine isethionate) in the treatment of kala-azar occurring in North Bihar, India. *Trans R Soc Trop Med Hyg*, **77**, 167–170.
23. Jha SN, Singh NKP, Jha TK (1991). Changing response to diamidine compounds in cases of kala-azar unresponsive to antimonial. J Assoc Physicians Indian, **39**, 314–316.
24. Thakur CP, Kumar M, Pandey AK (1991). Comparison of regimens of treatment of antimony-resistant Kala-azar patients: A randomized study. *Am J Trop Med Hyg*, **45**, 435–441.
25. Soto-Mancipe J, Grogl M, Berman J (1992). Evaluation of pentamidine for the treatment of cutaneous leishmaniasis in Colombia. *Clin Infect Dis*, **16**, 417–425.
26. Mishara M, Biswas UK, Jha DN, Khan AB (1992). Amphotericin versus pentamidine in anti-mony-unresponsive kala-azar. *Lancet*, **340**, 1256–1257.
27. Pentamidine. *Therapeutic Drugs*, edited by Sir Colin Dollery (1990), (London: Churchill Livingstone), pp. P20–P24.
28. Navin TR, Fontaine RE (1984). Intravenous versus intramuscular administration of pentamidine. *N Engl J Med*, **311**, 1701–1702.
29. Murphey SA, Josephs AS (1981). Acute pancreatitis associated with pentamidine therapy. *Arch Intern Med*, **141**, 56–58.
30. Alston TA (1988). Inhibition of cholinesterase by pentamidine. *Lancet*, **ii**, 1423.
31. *WHO Model Prescribing Information. Drugs used in parasitic diseases* (1990). (Geneva: World Health Organization).

Piperazine

Chemical structure

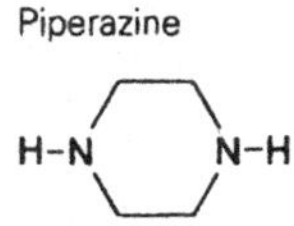

Physical properties

- Piperazine base (anhydrous): MW 86; pKa: 5.6, 9.8.
- Piperazine hexahydrate: MW 194. Freely soluble in water.
- Piperazine adipate: MW 232. 1 g dissolves in 18 ml of water.
- Piperazine phosphate: MW 202. 1 g dissolves in 60 ml of water.
- Tripiperazine dicitrate (piperazine citrate): MW 643. 1 g dissolves in 1.5 ml of water.

Pharmacology and mechanism of action

Piperazine is a heterocyclic organic base widely used as an anthelminthic. It was originally developed for the treatment of gout. Its first successful use in helminthiasis was reported by Mouriquand *et al.* in 1951 (1). Presently the drug is used in the treatment of infections caused by *Ascaris lumbricoides* and *Enterobius vermicularis*.

The drug causes flaccid paralysis in susceptible worms and the parasites lose their attachment to the intestinal wall, and are swept away by the normal bowel peristalsis. The biochemical mechanism behind this action is uncertain. Piperazine causes hyperpolarization of the *Ascaris* muscle rendering it unresponsive to acetylcholine (2).

Pharmacokinetics

GC (3, 4) and GC/MS (5) methods have been described for the determination of the drug.

No data on bioavailability have been reported. Early pharmacokinetic studies based on unspecific colorimetric methods have reported the drug to be quickly absorbed and eliminated with large inter-individual variability (6, 7). Using a GC method Fletcher *et al.* (3) studied the urinary excretion of piperazine in 34 healthy volunteers given 2.25 g of the drug as hexahydrate in 15 ml elixir. The 24 hour urinary excretion rate of piperazine varied between 8 and 30% of the orally administered dose. No metabolites were found in the urine with the method used and there was no evidence to show that the variation was due to differences in metabolism (3). Using a specific GC method, Tricker *et al.* (4) have recently reported similar findings. The 24 hour urinary excretion rate in 14 healthy volunteers given 2 g of piperazine citrate in 30 ml syrup varied between 11 and 59% (4).

Clinical trials

The effectiveness of piperazine in *Enterobius vermicularis* was first reported by Mouriquand *et al.* (1). A series of open studies conducted later confirmed those earlier findings (8–11).

Piperazine citrate was given at a dosage schedule of around 36 mg/kg once daily for both adults and children for two 1-week periods separated by a week of rest. The overall cure rate of the studies varied between 85 and 97%. Side effects such as urticaria, nausea or diarrhoea were rarely seen. In another study, Brown and Chan (12) treated 60 patients with a similar dose of piperazine citrate as above with a single dose given before breakfast on 7 consecutive days. The authors reported no advantages in prolonging the treatment to 14 days or dividing the daily dose.

The efficacy of piperazine in ascariasis was first reported in 1949 by Fayard (13). The author reported that thousands of French patients were treated with the drug with success rates between 70 and 95%. The dose used was 1.5 g of piperazine given in the evening and in the morning. Using a dose of 36 mg/kg of piperazine, Brown and Sterman (14) reported cure rates of 92% after 5 days of therapy; 87% after 4 days of therapy; 93% after 3 days of therapy; and 85% after 2 days of therapy. In another study (12), the same author treated 99 patients with ascariasis with a single dose of 50–75 mg/kg piperazine, and reported a cure rate of 74% after 1-day treatment, and 94% after 2-day treatment.

Indications

Treatment of infections due to *Ascaris lumbricoides* and *Enterobius vermicularis*. When cost and availability are not a consideration, safer and more effective drugs such as mebendazole or albendazole should be used instead.

Pregnancy and lactation

Piperazine has been used during pregnancy without teratogenic effects (15, 16), but there is a report of two infants with malformations, whose mothers apparently had taken the drug during pregnancy, although no causal relationship could be established (17). Because of its possible effect on the central nervous system and hypersensitivity reactions, piperazine should only be used if there is a strong justification for its use.

Its excretion into breast milk is unknown.

Side effects

Side effects commonly encountered with the recommended doses of piperazine are nausea, vomiting, abdominal cramps and diarrhoea which are usually mild and self-limiting. Although absolute incidence is unknown, severe side effects reported in the literature are rare. They can be classified into:

1. Allergic reactions such as urticaria, exantema, hypersensitivity, lacrimation, rhinorrea, productive cough, and bronchospasm (18, 19).
2. Neuro-psychological reactions (20–26):
   (a) cerebral type such as vertigo, dizziness, tremor, incoordination, ataxia and hypotonia with EEG changes;
   (b) psychic type such as depersonalization, hallucination and paranoic reactions;
   (c) miscellaneous such as headache, visual disturbances, somnolence, coma and an increase in the number of petit mal attacks.

Neuro-psychological reactions are rare. Most cases reported concern children with predisposing factors like neurological symptoms, renal diseases or those who have been treated with high doses of piperazine.

One case of haemolytic anaemia in a patient with G6PD deficiency (27), and one case of toxic hepatitis (28) have also been reported. However, no causal relationships can be established from these cases.

Nitrosation of piperazine to the potential carcinogen N-mononitrosopiperazine in the stomach of patients treated with normal therapeutic doses has been reported (4, 29). However, carcinogenicity related to the use of piperazine has not been reported despite the use of the drug over many years. In any case, this is unlikely to have any clinical implications with the short treatment period of nematodes.

Contraindications and precautions

Piperazine should not be given to patients with hypersensitivity or with neurological diseases, especially epileptic patients.

Interactions

In rats and mice, piperazine 1–5 g/kg subcutaneously, potentiates the side effects of chlorpromazine (30). However, this is unlikely to have any clinical significance. Piperazine is antagonistic to pyrantel, bephenium and levamisole (2), but no potential clinical interactions have been reported.

Dosage (31)

Note that the dosage of piperazine is often expressed as piperazine hexahydrate.

Infections caused by Ascaris lumbricoides

Adults
A single dose of 75 mg/kg of piperazine hexahydrate (to a maximum of 3.5 g).
Children
A single dose of 50 mg/kg of piperazine hexahydrate (to a maximum of 2.5 g).

Infections caused by Enterobius vermicularis (oxyuriasis)

Adults and children
A single dose of 50 mg/kg of piperazine hexahydrate daily for 7 days. This course is repeated after an interval of 2–4 weeks.

100 of anhydrous piperazine base is approximately equivalent to:
270 mg of piperazine adipate.
226 mg of piperazine hexahydrate.
235 mg of piperazine phosphate.
249 mg of tripiperazine dicitrate.

Preparations

Several preparations, apart from the one mentioned below, containing various piperazine salts are available.

- Antepar® (Wellcome). Oral suspension 150 mg piperazine hexahydrate/ml. Tablets 500 mg piperazine hexahydrate.

References

1. Mouriquand G, Roman E, Coisnard J (1951). Essai de traitement de l'oxyurose par la piperazine. *J Méd Lyon*, **32**, 189–195.
2. del Castillo J, De Mello WC, Morales T (1964). Mechanism of the paralysing action of piperazine on *Ascaris* muscle. *Br J Pharmacol*, **22**, 463–477.
3. Fletcher KA, Evans DAP, Kelly JA (1982). Urinary piperazine excretion in healthy Caucasians. *Ann Trop Med Parasitol*, **76**, 77–82.
4. Tricker AR, Kumar R, Siddiqi M, Khuroo MS, Preussmann R (1991). Endogenous formation of N-nitrosamines from piperazine and their urinary excretion following anthelminthic treatment piperazine citrate. *Carcinogenesis*, **12**, 1595–1599.
5. Skarping G, Bellander T (1986). Determination of piperazine in working atmosphere and in human urine using derivatization and capillary gas chromatography with nitrogen and mass selective detection. *J Chromatogr*, **370**, 245–258.
6. Rogers EW (1958). Excretion of piperazine salts in urine. *BMJ*, **I**, 136–137.
7. Hanna S, Tang A (1973). Human urinary excretion of piperazine citrate from syrup formulations. *J Pharm Sci*, **62**, 2024–2025.
8. White RHR, Standen OD (1953). Piperazine in the treatment of Threadworms in children: Report on a clinical trial. *BMJ*, **2**, 755–757.
9. Bumbalo TS, Gustina FJ, Oleksiak R (1954). The treatment of Pinworm infection (enterobiasis). *J Pediat*, **44**, 386–391.
10. Brown HW, Chan KF (1955). The treatment of *enterobius vermicularis* infections with piperazine. *Am J Trop Med*, **4**, 321–325.
11 Rachelson MH, Ferguson WR (1955). Piperazine in the treatment of enterobiasis. *Am J Dis Child*, **89**, 346–349.
12. Brown HW, Chan KF, Hussey K (1956). Treatment of enterobiasis and ascariasis with piperazine. *J Am Med Ass*, **161**, 515–520.
13. Fayard C (1949). Ascaridiose et piperazine. Thesis, Faculté de Médécine de Paris.
14. Brown H, Sterman MM (1954). Treatment of *Ascaris lumbricoides* infection with piperazine citrate. *Am J Trop Med*, **3**, 750–754.
15. Heinonen OP, Slone D, Shapiro S (1977). *Birth defects and drugs in pregnancy*. (Littletown, Massachusetts: PSG), 297.
16. Villar AAL, Sibai B (1992). Nematode infections: Is it wise to withhold medical treatment during pregnancy? 1. Biliary implications. *Am J Obstet Gynecol*, **166**, 549–550.
17. Leach FN (1990). Management of Threadworm infestations during pregnancy. *Arch Dis Child*, **65**, 399–400.
18. Macmillan AL (1973). Generalized pustular drug rash. *Dermatologia*, **146**, 285–291.
19. McCullagh SF (1968). Allergenicity of piperazine: a study in environmental aetiology. *Br J Ind Med*, **25**, 319–325.
20. Belloni C, Rizzoni G (1967). Neurotoxic side-effects of piperazine. *Lancet*, **ii**, 369.
21. Berger JR, Globus M, Melamed E (1979). Acute transitory cerebellar dysfunction associated with piperazine adipate. *Arch Neurol*, **36**, 180–181.
22. Bomb RS, Bedi HK (1976). Neurotoxic side-effects of piperazine. *Trans R Soc Trop Med Hyg*, **70**, 358.
23. Gupta SR (1976). Piperazine neurotoxicity and psychological reaction. *J Ind Med Ass*, **66**, 33–34.
24. Parsons AC (1971). Piperazine neurotoxicity. 'Worm wobble'. *BMJ*, **4**, 790–792.
25. Vallat JN, Vallat JM, Texier J, Léger J (1972). Les signes neurologiques d'intoxication par la pipérazine. *Bordeaux Médicale*, **5**, 394–400.
26. Nickey LN (1966). Possible precipitation of petit mal seizures with piperazine citrate. *J Am Med Ass*, **195**, 193–194.
27. Buchanan N, Cassel R, Jenkins T (1971). G-6-PD deficiency and piperazine. *BMJ*, **2**, 110.
28. Hamlyn AN, Morris JS, Sarkany I, Sherlock S (1976). Piperazine hepatitis. *Gastroenterology*, **70**, 1144–1147.

29. Bellander T, Österdahl B-G, Hagmar L (1985). Formation of N-mononitrosopiperazine in the stomach and its excretion in the urine after oral intake of piperazine. *Toxicol Appl Pharmacol*, **80**, 193–198.
30. Sturman G (1973). Interaction between piperazine and chlorpromazine. *Br J Pharmacol*, **50**, 153–155.
31. *WHO Model Prescribing Information. Drugs used in parasitic diseases* (1990), (Geneva: World Health Organization).

Praziquantel

Chemical structure

Praziquantel

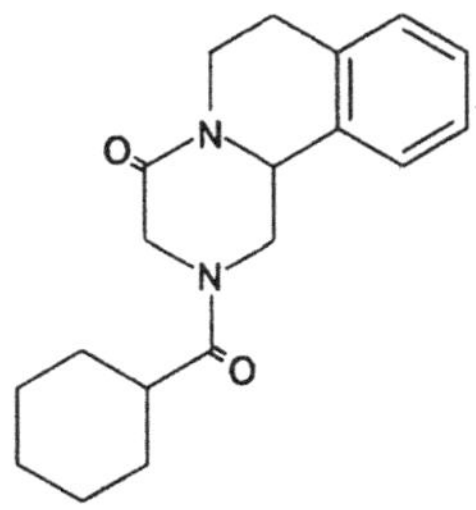

Physical properties

MW 312. Neutral compound. Slightly soluble in water. The drug should be protected from light.

Pharmacology and mechanism of action

Praziquantel is a pyrazinoquinoline compound originally developed for the treatment of schistosomiasis but has been found to have a wide spectrum of anthelminthic activity. Praziquantel is a racemate but the R (+) enantiomer is solely responsible for its antiparasitic activity. It is active against trematodes (all *Schistosoma* species pathogenic to man, *Paragonimus westermani*, and *Clonorchis sinensis*) and cestodes (*Taenia saginata*, *Taenia solium*, *Hymenolepis nana* and *Diphyllobothrium latum*) (1).

The mechanism of action of praziquantel is not clearly known. Schistosomes take up the drug rapidly. Drug uptake is immediately followed by increased muscular activity that proceeds to tetanic contraction and vacuolization of the parasite tegument (2). The muscular effects of the drug are presumed to be responsible for the shift of the parasites from the mesenteric veins to the liver *in vivo*. However, hepatic shift has been demonstrated with most known schistosomicides and may not provide any specific information of the drug's mechanism of action. Recent experimental findings have suggested that the anti-schistosomal effects of the drug are related to its effect on the tegument rather than on the musculature (3). Another pharmacological effect of the drug includes an increase of membrane permeability to cations, particularly calcium (4). However, the role of this effect to the anthelminthic property of the drug is unknown.

Pharmacokinetics

Specific HPLC (5, 6) and GC (7) analytical methods have been described for the determination of praziquantel and its metabolites.

Absolute oral bioavailability is unknown. However, it is presumed that more than 80% of the drug is absorbed, although this may vary between the different commercial preparations (8, 9). The absorption of the drug has been reported to be dose dependent (10, 11). In one study, a four-fold increase in oral dose led to an eight-fold rise in serum concentrations of the drug which could be due to a first pass effect (11). Food has been reported to increase its bioavailability (8). Peak plasma levels are reached after 1–3 hours. Plasma levels of the parent drug are very low and vary widely between individuals (9–11). Praziquantel has an apparent volume of distribution of about 700 l (12). It crosses the blood–brain barrier attaining a concentration in CSF of about 25% of that in plasma (13). It is extensively metabolized to several mono- and poly-hydroxylated metabolic products (14). The metabolites may have some pharmacological activities (1). The plasma elimination half-life for the unchanged drug is 1 to 1.5 hours and 4 to 5 hours for the metabolites after a single oral dose. Within 24 hours more than 90% of the dose was recovered in the urine as the mono- and poly-hydroxylated metabolites and less than 1% as the parent drug (11, 14, 15).

Clinical trials

In well designed multicentre clinical trials undertaken in Africa (16–18), Asia (19–21) and Latin America (22), the use of praziquantel in the treatment of schistosomiasis (*haematobium*, *mansoni* and *japonicum*) showed cure rates of 75–95% after 1–6 months. The drug was given either as a single dose of 20–50 mg/kg, or 20 mg/kg once daily for 2 days, or 25 mg/kg once daily for 3 days. Similar efficacy was reported in trials on *Schistosoma mekongi* (23). The overall results show that a single oral dose of 40 mg/kg for 1 day gives a cure rate of 95% in *S. haematobium* infection and 90% in *S. mansoni* infection. Cure rates of around 70% may be obtained after a higher dose of 30 mg/kg daily for 2 days against *S. japonicum* (1). In two recent studies it has been shown that smaller doses of 20–30 mg/kg for 1 day (24, 25) could give similar results as for the 40 mg/kg for 1 day used against *Schistosoma haematobium*. Less curative doses of the drug may, however, lead to the appearance of resistant schistosomes because this possibility has been shown experimentally in the laboratory.

Single doses of 10 mg/kg given to children and adults infected with *T. saginata* and *T. solium* gave cure rates close to 100% (26). A higher dose (25 mg/kg) was required to achieve similar results in infections with *Hymenolepis nana* and *Diphyllobothrium latum* (27, 28).

Although data are still not conclusive, praziquantel has been shown to be effective against certain liver flukes (*Clonorchis*/*Opistorchis*), lung flukes (*Paragonimus*) and against the larval stages of *Taenia solium* (cysticercosis) (2).

Indications

Infections caused by *Schistosoma* species pathogenic to man (*Schistosoma haematobium*, *S. mansoni*, *S. japonicum* and *S. mekongi*). The drug is most cost-effective in mixed infections. It is also effective for infections with flukes (*Paragonimus westermani* and *Clonorchis sinensis*) and in cestodes (*Hymenolepis nana*, *Diphyllobotrium latum*, *Taenia saginata*, *T. solium*) including the larval stage of *Taenia solium* (cysticercosis).

Praziquantel has some effect against fascioliasis, but triclabendazole, a new anthelminthic drug still under clinical evaluation is more effective.

Pregnancy and lactation

Teratogenicity has not been reported in rats (29). Documentation in man is lacking, despite the widespread use of the drug. Treatment with praziquantel should be postponed until delivery, unless there is a strong indication for its use.

Clinically insignificant amounts of the drug are excreted into breast milk (30).

Side effects

In large-scale and community-based studies in patients and healthy volunteers, the drug showed only mild to moderate and transient side effects (16–22). The frequency and intensity of side effects seemed to be dose related. In one study (21), the frequency of the side effects were: dizziness (29%), headache (15%), lassitude (19%), pain in the limbs (22%), and abdominal distress (9%). Nausea, insomnia, fever, and non-itching macular eruptions occurred in single patients. 40% of the patients remained free from any side effects.

Abdominal colic and bloody diarrhoea due to praziquantel have been reported by others (32, 33). Praziquantel has not shown to be mutagenic or carcinogenic (28, 30, 31).

Contraindications

Dosage has to be reduced in patients with liver diseases.

Interactions

Phenytoin, carbamazepine, and dexamethasone have been reported to decrease the plasma concentrations of praziquantel by 10% to 50% (34, 35). The clinical relevance of these interactions for the treatment of parasitic infections needs further investigation.

Dosages

Schistosoma haematobium and S. mansoni

40 mg/kg as a single dose.

Schistosoma japonicum, S. intercalatum, and S. mekongi

60 mg/kg divided into 2 doses given in 1 day. Experience in the treatment of *S. intercalatum* is limited.

Diphyllobotrium latum and Hymenolepis nana

25 mg/kg as a single dose.

Taenia saginata and T. solium

10–20 mg/kg as a single dose.

Flukes (intestinal, liver and lungs)

25 mg/kg three times daily for 1–2 days.

Preparations

- Biltricide® (Bayer). Tablets 600 mg.
- Cysticide® (E. Merck). Tablets 500 mg.
- Cesol® (E. Merck). Tablets 150 mg.

References

1. Andrews P, Thomas H, Pohlke R, Seubert J. Praziquantel (1983). *Med Res Rev*, **3**, 147–200.
2. Xiao SH, Friedman PA, Catto BA, Webster LT Jr (1984). Praziquantel induced vesicle formation in the tegument of male *mansoni* is calcium dependent. *J Parasitol*, **70**, 177–179.
3. Xiao SH, Catto BA, Webster LT Jr, Melborn H, Becker B (1984). Effects of praziquantel on different developmental stages of *Schistosoma mansoni in vitro* and *in vivo*. *J Infect Dis*, **151**, 1130–1137.
4. Pax R, Bennett JL, Fetterer R (1978). A benzodiazepine derivative and praziquantel: effects on musculature of *Schistosoma mansoni* and *Schistosoma japonicum*. *Naunyn Schmiedebergs Arch Pharmacol*, **304**, 309–315.
5. Westhoff F, Blaschke F (1992). High-performance liquid chromatographic determination of the stereoselective biotransformation of the chiral drug praziquantel. *J Chromatogr*, **578**, 265–271.
6. Gonzales-Esquivel DF, Okuno CM, Sanchez RM, Solelo MJ, Cook HJ (1993). Sensitive high-performance liquid chromatographic assay for praziquantel in plasma, urine and liver homogenates. *J Chromatogr*, **613**, 174–178.
7. Diekmann HW (1979). Quantitative determination of praziquantel in body fluids by gas-chromatography. *Eur J Drug Metab Pharmacokinet*, **4**, 139–141.
8. Mandour El M, Turabi EL H, Homeida MA, Sadig EL T, Ali HM, Bennet JL, Leahey WJ, Harron WG (1990). Pharmacokinetics of praziquantel in healthy volunteers and patients with schistosomiasis. *Trans R Soc Trop Med Hyg*, **84**, 389–393.
9. Kaojarern S, Nathakarnkikool S, Suvanakoot U (1989). Comparative bioavailability of praziquantel tablets. *DICP Ann Pharmacother*, **23**, 29–32.
10. Leopold G, Ungethum W, Groll E, Diekmann HW, Nowak H, Wegner DHG (1978). Clinical pharmacology in normal volunteers of praziquantel, a new drug against schistosomes and cestodes. *Eur J Clin Pharmacol*, **14**, 281–291.
11. Bittencourt PRM, Gracia CM, Gorz AM, Oliveira TV (1990). High-dose praziquantel for neurocysticercosis: serum and CSF concentrations. *Acta Neurol Scand*, **82**, 28–33.
12. Jung H, Vazquez ML, Sanchez M, Penagos P, Sotelo J (1991). Clinical pharmacokinetics of praziquantel. *Proc West Pharmacol Soc*, **34**, 335–340.
13. Jung H, Hurtado M, Sanchez M, Medina MT, Sotelo J (1990). Plasma and CSF levels of albendazole and praziquantel in patients with neurocysticercosis. *Clin Neuropharmacol*, **13**, 559–564.
14. Bühring KU, Diekmann HW, Müller H, Garbe A, Nowak H (1978). Metabolism of praziquantel in man. *Eur J Drug Metab Pharmacokinet*, **3**, 179–190.
15. Patzschke K, Pütter J, Wegneu LA, Horster FA, Diekmann HW (1979). Serum concentrations and renal excretion in humans after oral administration of praziquantel: results of three determination methods. *Eur J Drug Metab Pharmacokinet*, **4**, 149–156.
16. Davis A, Biles JE, Ulrich AM, Dixon H (1981). Tolerance and efficacy of praziquantel in phase IIA and IIB therapeutic trials in Zambian patients. *Arzneimittelforschung*, **31**, 568–574.
17. Davis A, Biles JE, Ulrich AM (1979). Initial experiences in patients with *Schistosoma mansoni* previously treated with oxamniquine and/or hycanthone: Resistance of *Schistosoma mansoni* to schistosomicidal agents. *Trans R Soc Trop Med Hyg*, **76**, 652–659.
18. Pugh RNH, Teesdale CH (1983). Single dose oral treatment in urinary schistosomiasis: a double blind trial. *BMJ*, **286**, 429–432.
19. Ishizaki T, Kamo E, Boehme K (1979). Double-blind studies of tolerance to Praziquantel in Japanese patients with *Schistosoma japonicum* infections. *Bull WHO*, **57**, 787–791.
20. Santos AT, Blas BL, Nosenas JS, Portillo GP, Ortega OM, Hayashi M, Boehme K (1979). Preliminary clinical trials with praziquantel in *Schistosoma japonicum* infections in the Philippines. *Bull WHO*, **57**, 793–799.
21. Zhejiang Clinical Cooperative Research Group for praziquantel (1980). Clinical evaluation of

praziquantel in treatment of schistosomiasis japonica. A report of 181 cases. *Chin Med J*, **93**, 375–384.

22. Katz N, Rocha RS, Chaves A (1979). Preliminary trial with praziquantel in human infections due to *Schistosoma mansoni*. *Bull WHO*, **57**, 781–785.
23. Nash TE, Hofstetter M, Cheever AW, Ottesen EA (1982). Treatment of *Schistosoma mekongi* with praziquantel: a double-blind study. *Am J Trop Med Hyg*, **31**, 977–982.
24. King CH, Wiper DW, De Stiger KV, Peters PAS, Koech D, Ouma JH, Arap Siongok TK, Mahamoud AAF (1989). Dose-finding study for praziquantel therapy of *Schistosoma haematobium* in coast province, Kenya. *Am J Trop Med Hyg*, **40**, 507–513.
25. Mott KE, Dixon H, Osei-Tutu E, England EC, Davis A (1985). Effect of praziquantel on haematuria and proteinuria in urinary schistosomiasis. *Am J Trop Med Hyg*, **34**, 1119–1126.
26. Gemmell MA, Johnstone PD (1981). Cestodes. *Antibiot Chemother*, **30**, 54–114.
27. Bylund G, Bång B, Wikgren K (1977). Tests with a new compound (Praziquantel) against *Diphyllobotrium latum*. *J. Helminthol*, **51**, 115–119.
28. Frohberg H, Schulze Schenking M (1981). Toxicological profile of praziquantel a new drug against cestode and *Schistosoma* infections as compared to some other schistosomicides. *Arzneimittelforschung*, **31**, 555–565.
29. Ni YC, Shao BR, Zhan CQ, Xu YQ, Ha SH, Jiao PY (1982). Mutagenic and teratogenic effects of anti-schistosomal praziquantel. *Chin Med J (Engl)*, **95**, 494–498.
30. Pütter J, Held H (1979). Quantitative studies on the occurrence of praziquantel in milk and plasma of lactating women. *Eur J Drug Metab Pharmacokinet*, **4**, 193–198.
31. Billings PC, Heidelberger C (1982). Effects of praziquantel a new antischistosomicide drug on the mutation and transformation of mamalian cells. *Cancer Res*, **42**, 2692–2696.
32. Watt G, Baldovino P, Castro J, Fernando M, Ranoa C (1986). Bloody diarrhea after praziquantel therapy. *Trans R Soc Trop Med Hyg*, **80**, 345–346.
33. Polderman AM, Gryseels B, Gerold JL, Mpamila K, Manshande JP (1984). Side effects of praziquantel in the treatment of *Schistosoma mansoni* in Maniema, Zaire. Trans R Soc Trop Med Hyg, **78**, 752–754.
34. Bittencourt PRM, Gracia CM, Martins R, Fernandes AG, Diekmann HW, Jung W (1992). Phenytoin and carbamazepine decrease oral bioavailability of praziquantel. *Neurology*, **42**, 492–496.
35. Vazquez M, Jung H, Sotelo J (1987). Plasma levels of praziquantel decrease when dexamethasone is given simultaneously. *Neurology*, **37**, 1561–1562.

Primaquine

Chemical structure

Primaquine

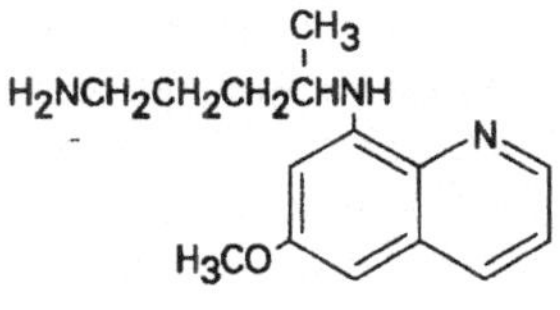

Physical properties

Base: MW 259; diphosphate: MW 455; pKa is not known. 1 g dissolves in 16 ml water. It should be protected from light.

Pharmacology and mechanism of action

Primaquine is active against primary exoerythrocytic stages of all malaria parasites. Primaquine is also effective against latent exoerythrocytic stages of *P. vivax* and *P. ovale* responsible for relapse. It possesses gametocytocidal activity against all four species of plasmodia which infect man and could theoretically be used to block malaria transmission. Primaquine has no effect on the erythrocytic stages of plasmodia unless toxic concentrations are achieved (1).

The mechanism of action of primaquine is unknown.

Pharmacokinetics

Specific HPLC (2), GC/MS (3), and GC (4) analytical methods have been described for primaquine and its metabolites.

Primaquine is rapidly and completely absorbed with an absolute oral bioavailability of around 96% (5). Peak plasma levels of 150–200 ng/ml are reached within 2–3 hours after a 45 mg dose. It is extensively distributed into body tissue with a volume of distribution of around 3–4 l/kg (5–8). Primaquine is a low clearance drug (24.2± 7.4 l/h) and is mainly eliminated by metabolism. According to one study using ^{14}C-primaquine, the total radioactivity in the plasma declined slowly and was still significant after 4 days. Primaquine itself was almost completely eliminated during the first 24 hours, with a plasma half-life of around 7 hours (5). Only 1% of an oral dose was excreted unchanged through the kidneys over 24 hours (6–8). One major metabolite, carboxyprimaquine, has been identified in human blood (7) and another, 6-methoxyprimaquine, in the urine in small amounts (2). Carboxyprimaquine accumulates in the plasma to much higher concentrations than that of the parent compound because of its slow elimination from the body (5, 9).

Repeated dosing had no effect on the pharmacokinetic parameters in healthy Thai volunteers and the values were in broad agreement with those obtained in Caucasians (10).

Indications

Primaquine is used for the radical treatment of *P. vivax* and *P. ovale* infections. It has also been used in large populations to avoid spread of chloroquine-resistant strains of *falciparum* malaria (through its gametocytocidal action).

Clinical trials

There are no recent clinical trials of the efficacy of primaquine and therefore the actual susceptibility for the drug is uncertain in most parts of the world. It is well known that strains of *P. vivax* vary in their response to primaquine. Adults infected with the Chesson type of strain (reported from New Guinea, Solomon islands, Indonesia, Thailand) should therefore receive twice the standard dose to prevent relapse (1, 11). The standard treatment is, however, sometimes ineffective in preventing relapse of *P. vivax* also in other parts of the world (12).

Side effects

Primaquine is usually well tolerated in the therapeutic dosage of 15 mg base/day for 14 days, but abdominal pain and gastric distress are common if administered on an empty stomach (13). The severity of the gastrointestinal side effects are dose-related, and with larger doses, nausea and vomiting occurs. Rare effects include hypertension and cardiac arrhythmia (13).

The principal toxic effect of primaquine is haemolytic anaemia especially in patients with a deficiency of glucose-6-phosphate dehydrogenase (G6PD)(1) It is estimated that 200–300 million people have G6PD deficiency (14). The acute haemolysis is seen after a latent phase of 1–3 days. Drug administration should be discontinued when a darkening of the urine (if possible check urine urobilinogen after 1–3 days) or a sudden decrease in haemoglobin levels occurs. The prognosis of this condition is good and specific treatment usually not needed (1). In addition to haemolytic anaemia, primaquine can also cause methaemoglobinaemia and may rarely suppress bone marrow activity leading to leucopenia (13, 14).

Pregnancy and lactation

Documentation on teratogenicity is lacking both in animals and in man. However, primaquine is contraindicated during pregnancy because of the possibility that it passes the placenta and may cause haemolytic anaemia in a G6PD deficient fetus (15).

The drug passes into breast milk, and mothers taking it should not breast-feed (14).

Contraindications and precautions

It is recommended not to use primaquine in patients with conditions affecting bone marrow function or on myelosuppressive medication (14).

Interactions

In one study, the effect of primaquine has been studied on the metabolism of antipyrine. Primaquine (45 mg) given 2 hours before antipyrine (300 mg orally), increased antipyrine half-life (calculated from 0 to 24 hours) from a mean of 13 to 25 hours and decreased clearance from 3 to 1 l/h (16).

Dosage

Adults
15 mg primaquine base (or 0.25 mg/kg) daily for 14 days following standard chloroquine therapy. For patients with Chesson strains (see Clinical trials), higher doses of primaquine (30 mg base) may be required (1). When used as a gametocytocide in *falciparum* malaria, a single dose of 30–40 mg base is given (17).
Children over 1 year
0.25 mg base/kg for 14 days after standard chloroquine therapy (17).

Preparations

Available as primaquine phosphate: 26.3 mg phosphate equals 15 mg base.

- Primaquine® (Zeneca) Tablets 13.2 mg.

References

1. Black RH, Canfield CJ, Clyde DF, Peters W, Wernsdorfer WH (1986). Primaquine. In: *Chemotherapy of Malaria*, 2nd edn, edited by L. Bruce-Chwatt. Geneva: World Health Organization, pp. 61–63.
2. Ward SA, Edwards G, Orme ML'E, Breckenridge AM (1984). Determination of primaquine in biological fluids by reversed-phase high performance liquid chromatography. *J Chromatogr*, **305**, 239–243.
3. Greaves J, Evans DAP, Gilles HM, Baty JD (1979). A selected ion monitoring assay for primaquine in plasma and urine. *Biomed Mass Spectrometr*, **6**, 109–112.
4. Rajagopalan TG, Anjaneyula B, Shanbag VD, Grewal RS (1981). Electron capture gas chromatography assay for primaquine in blood. *J Chromatogr*, **224**, 265–273.
5. Mihaly GW, Ward SA, Edwards G, Nicholl DD, Orme ML'E, Breckenridge M (1985). Pharmacokinetics of primaquine in man. I. Studies of the absolute bioavailability and effects of dose size. *Br J Clin Pharmacol*, **19**, 745–750.
6. Greaves J, Evans DAP, Gilles HM, Fletcher KA, Bunnag D, Harinasuta T (1980). Plasma kinetics and urinary excretion of primaquine in man. *Br J Clin. Pharmacol*, **10**, 399–405.
7. Mihaly GW, Ward SA, Edwards G, Nicholl DD, Orme ML'E, Breckenridge AM (1984). Pharmacokinetics of primaquine in man: identification of the carboxylic acid derivative as a plasma metabolite. *Br J Clin Pharmacol*, **17**, 441–446.
8. Fletcher KA, Price-Evans DA, Gilles HM, Greaves J, Bunnag D (1981). Studies of the pharmacokinetics of primaquine. *Bull World Health Organ*, **59**, 407–412.
9. Bhatia SC, Saraph YS, Revankar SN, Doshi KJ, Bharucha ED, Desai ND, Vaidya AB, Subramanyam D, Gupta KC, Satoskar RS (1986). Pharmacokinetics of primaquine in patients with *P. vivax* malaria. *Eur J Clin Pharmacol*, **31**, 205–210.
10. Ward SA, Mihaly GW, Edwards G, Looareesuwan S, Phillips RE, Chanthavanich P, Warrel DA, Orme ML'E, Breckenridge AM (1985). Pharmacokinetics of primaquine in man. II: Comparison of acute *vs* chronic dosage in Thai subjects. *Br J Clin Pharmacol*, **19**, 751–755.
11. Coatney GR, Getz ME (1962). Primaquine and quinocide as curative against sporozoite-induced Chesson strain *vivax* malaria. *Bull World Health Organ*, **27**, 290.
12. Rombo L, Edwards G, Eriksson G, Lindquist L, Lindberg A, Runehagen A, Ward SA, Björkman A, Hylander NO (1987). Seven patients with relapses of *P. vivax* and *P. ovale* despite primaquine treatment. *Trop Med Parasit*, **38**, 49–50.

13. Primaquine. *Martindale: The Extra Pharmacopoeia*, 28th edn (1982). (London: Pharmaceutical Press), p. 404.
14. Primaquine. *Therapeutic Drugs*, edited by Sir Colin Dollery (1991). (London: Churchill Livingstone), pp. P209–P213.
15. Centers for Disease Control, (1990). Recommendations for the prevention of malaria among travellers. *Morbidity and Mortality Weekly Report*, **39**, (No-RR-3), 1–10.
16. Back DJ, Purba HS, Park BK, Ward SA, Orme ML'E (1983). Effect of chloroquine and primaquine on antipyrine metabolism. *Br J Clin Pharmacol*, **16**, 497–502.
17. *WHO Model Prescribing Information. Drugs used in parasitic diseases* (1990). (Geneva: World Health Organization).

Proguanil

Chemical structure

Proguanil

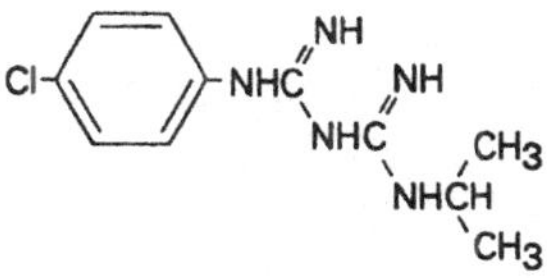

Physical properties

Base: MW 254; hydrochloride: MW 290; pKa: 2.3, 10.4. 1 g of the salt dissolves in 110 ml of water. The drug should be protected from light.

Pharmacology and mechanism of action

Proguanil (PG) was introduced as a prophylactic agent against malaria just after the Second World War. It is a pyrimidine derivative which is highly active against pre-erythrocytic forms of *Plasmodium (P) falciparum* making it suitable for casual prophylaxis. It is also effective in the erythrocytic phase (schizontocide) against all forms of malaria, but the action is too slow for the drug to be used for treatment. Proguanil prevents the formation of sporozoites in the mosquito, thus interfering with the spread of malaria (1). The drug acts through an active metabolite (cycloguanil). The mechanism of action is due to an inhibition of dihydropholate reductase (2).

Like most other antimalarials, the efficacy of proguanil has been reduced by the development of resistence. Already during the 1950s and 60s *P. falciparum* resistance was reported from all endemic areas including Africa (1). In *P. vivax* and *P. malariae*, resistence seems less frequent but resistent strains have been reported in Malaysia, Indonesia, and Taiwan (1). Partial cross-resistance occurs with other antifolates, particularly with pyrimethamine.

Pharmacokinetics

Specific HPLC methods have been described for determination of proguanil and its metabolites (3–6).

Proguanil is only available in oral formulation. It is rapidly absorbed but the absolute bioavailability is unknown. Using a specific analytical method (6), the pharmacokinetics of proguanil and its two metabolites was studied in adult healthy volunteers (7). After a single oral dose of 200 mg of proguanil hydrochloride, peak plasma levels of 150–220 ng/ml were reached in 2–3 hours. The corresponding concentrations of its metabolites, cycloguanil (CG) and 4-chlorophenylbiguanide (CPB), represented about 24% and 6% of the parent drug, respectively. The plasma concentration of PG is about 20% of that in whole blood but the concentration of the active metabolite, CG, is similar in both media. Plasma and whole blood concentrations of proguanil and its metabolites declined in parallel with terminal elimination half-lives of around 16 hours (7). The plasma protein binding of PG is approximately 75% (8). Earlier studies reported that the drug is eliminated largely unchanged with the urine (60%) but also as CG (30%), and as CPB (10%) (9).

Proguanil is metabolized in the liver to CG by the cytochrome P450IIC sub-family. There is a pronounced inter-subject variability in the capacity to metabolize PG (10). Poor metabolizers of PG to CG are also slow hydroxylators of mephenytoin (11)). The prevalence of slow hydroxylators of mephenytoin varies in different ethnic groups and is 3% in Caucasians (12), and about 20% in Orientals (13). In Kenyans, as many as 35% were considered poor metabolizers of PG (14). The large inter-individual differences in plasma concentrations of the active metabolite CG are probably of clinical importance, but so far no studies have confirmed reduced efficacy in poor metabolizers of PG.

Clinical trials

Already in 1946, it was reported that a single dose of 10–100 mg PG given 2–5 days after exposure to *P. falciparum* could prevent development of a subsequent infection (15). It was suggested that 100 mg should be given twice weekly, but once daily became the generally accepted regimen. In 1984, a study from Tanzania with Japanese subjects claimed that 200 mg PG once daily was more effective than 100 mg as monoprophylaxis, and after this study the manufacturer recommended the higher dose (16). The study was retrospective and the diagnosis of malaria was not verified. All subjects took 100 mg PG daily and 12 out of 13 had malaria breakthroughs. Approximately 20% of all Japanese are poor metabolizers of PG and it is therefore not probable that underdosing was the explanation for alleged inefficiency in this ethnic group.

Proguanil-resistant *P. falciparum* strains are present worldwide. The prophylactic efficacy of PG is generally insufficient in South East Asia and the drug should not be recommended in this part of the world (17). In Tropical Africa, resistance is much less common, and studies in children have demonstrated a rather good efficacy in combination with chloroquine in Cameroon (18) and as monoprophylaxis in Tanzania (19). A recent prospective study in Dutch travellers to Africa did not find any difference in efficacy between PG 100 or 200 mg once daily in combination with chloroquine or PG 200 mg daily as monoprophylaxis (20).

Indications

Proguanil is used in combination with chloroquine as chemoprophylaxis against *falciparum* malaria in areas with a low frequency of resistance, i.e. tropical Africa.

Pregnancy and lactation

Teratogenicity in man has never been reported, despite the widespread use of the drug over many years. Proguanil is generally regarded safe during pregnancy, but folate supplementation at suitable intervals may be required.

Both proguanil and CG are excreted into breast milk with concentrations similar to those in plasma but inadequate to ensure reliable protection of the infant (21).

Side effects

Proguanil is well tolerated in recommended doses and severe side effects are not reported in persons with a normal kidney function (21). Several reports of mouth ulceration due to proguanil have, however, been reported (22, 23), and mild epigastric discomfort may occur.

Contraindications and precautions

Dosage adjustments are necessary in patients with kidney failure (24).

Interactions

Chloroquine may increase the risk of mouth ulceration with proguanil (25).

Dosage

Proguanil should not be used for treatment of malaria. Dosage of proguanil is expressed as the hydrochloride.

Prophylaxis (26)

Adults
200 mg daily.
Children
For children several different dosage regimens are used but the World Health Organization recommends 3 mg/kg daily or:

<1 year	25 mg
1–4 years	50 mg
5–8 years	75 mg
9–12 years	100 mg
>12 years	200 mg

Administration of the drug is usually recommended for at least 4 weeks after leaving malarious areas. However, after re-evaluation of the efficacy of proguanil it has recently been recommended in Sweden to reduce the adult dose to 100 mg daily and to continue for only one week after departure from endemic areas (27).

Preparation

Available as proguanil hydrochloride: 100 mg hydrochloride equals 87 mg base.

- Paludrine® (Zeneca). Tablets 100 mg.

References

1. Black RH, Canfield CJ, Clyde DF, Peters W, Wernsdorfer WH (1986). Proguanil and proguanil analogues. In: *Chemotherapy of Malaria*, 2nd edn, edited by L. J. Bruce-Chwatt. (Geneva: World Health Organization), pp. 71–77, 110–111.
2. Ferone R, Burchall JJ, Hitchings GH (1969). *Plasmodium berghei* dihydrofolate reductase. Isolation properties and inhibition by antifolates. *Mol Pharmacol*, **5**, 45–59.
3. Moody RR, Selkirk AB, Taylor RB (1980). High-performance liquid chromatography of

proguanil, cycloguanil and 4-chlorophenylbiquanide using hydrophobic pairing ion and its application to serum assay. *J Chromatogr*, **182**, 359–367.
4. Edstein MD (1986). Simultaneous meaurement of proguanil and cycloguanil in human plasma by high performance liquid chromatography. *J Chromatogr*, **380**, 184–189.
5. Kelly JA, Fletcher KA (1986). High performance liquid chromatographic method for the determination of proguanil and cycloguanil in biological fluids. *J Chromatogr*, **381**, 464–471.
6. Taylor RB, Moody RR, Ochekpe NA (1987). Determination of proguanil and its metabolites cycloguanil and 4-chlorphenylbiguanide in plasma, whole blood, and urine by HPLC. *J Chromatogr*, **416**, 394–399.
7. Wattanagoon Y, Taylor RB, Moody RR, Ochekpe NA, Looareesuwan S, White NJ (1987). Single dose pharmacokinetics of proguanil and its metabolites in healthy subjects. *Br J Clin Pharmacol*, **24**, 775–780.
8. Ritschel WA, Hammer GV, Thompson GA (1978). Pharmacokinetics of antimalarials and proposals for dosage regimens. *Int J Clin Pharmacol Biopharm*, **16**, 395–401.
9. Smith CC, Ihrig J, Menne R (1961). Antimalarial activity and metabolism of biguanides. I. Metabolism of chloroguanide and chloroguanide triazine in Rhesus monkeys and man. *Am J Trop Med Hyg*, **10**, 694–703.
10. Ward SA, Watkins WM, Mberu E, Saunders JE, Koech DK, Gilles HM, Howells RE, Breckenridge AM (1989). Inter-subject variability in the metabolism of proguanil to the active metabolite cycloguanil in man. *Br J Clin Pharmacol*, **27**, 781–787.
11. Ward SA, Helsby NA, Skjelbo E, Brosen K, Gram LF, Breckenridge AM (1991). The activation of the biguanide antimalarial proguanil co-segregates with the mephenytoin oxidation polymorphism – a panel study. *Br J Clin Pharmacol*, **31**, 689–692.
12. Wedlund PJ, Aslanian WS, McAllister CB, Wilkinson GR, Branch RA (1984). Mephenytoin hydroxylation deficiency in Caucasians: Frequency of a new oxidative drug metabolism polymorphism. *Clin Pharmacol Ther*, **36**, 773–780.
13. Horai Y, Nakano M, Ishizaki T, Ishikawa K, Zhou H-H, Zhou B-J, Liao C-L, Zhang L-M (1989). Metoprolol and mephenytoin oxidation polymorphism in Far Eastern Oriental subjects: Japanese versus mainland Chinese. *Clin Pharmacol Ther*, **46**, 198–207.
14. Watkins WM, Mberu EK, Nevill CG, Ward SA, Breckenridge AM, Koech DK (1990). Variability in the metabolism of proguanil to the active metabolite cycloguanil in healthy Kenyan adults. *Trans R Soc Trop Med Hyg*, **84**, 492–495.
15. Fairley NH (1946). Researches on paludrine (M.4888) in malaria. An experimental investigation undertaken by the L.H.Q. Medical Research Unit (AIF), Cairns, Australia. *Trans R Soc Trop Med Hyg*, **40**, 106–162.
16. McLarty DG, Webber RH, Jaatinen M, Kihamia CH, Murru M, Kumano M, Aubert B, Magnuson LW (1984). Chemoprophylaxis of malaria in non-immune residents in Dar es Salaam, Tanzania. *Lancet*, **2**, 656–659.
17. Henderson A, Simon JW, Melia W (1986). Failure of malaria chemoprophylaxis with a proguanil-chloroquine combination in Papua New Guinea. *Trans R Soc Trop Med Hyg*, **80**, 838–840.
18. Gozal D, Fada G, Hengy C (1991). Long-term chloroquine-proguanil malaria prophylaxis in a nonimmune pediatric population. *J Pediatr*, **118**, 142–145.
19. Rooth I, Sinani HM, Björkman A (1991). Proguanil daily or chlorproguanil twice weekly are efficacious against *falciparum* malaria in a holoendemic area of Tanzania. *J Trop Med Hyg*, **94**, 45–59.
20. Wetsteyn JCFM, de Geus A (1993). Comparison of three regimens for malaria prophylaxis in travellers to east, central and southern Africa. *BMJ*, **307**, 1041–1043.
21. Proguanil. *Therapeutic Drugs*, edited by Sir Colin Dollery (1991), (London: Churchill Livingstone), pp. P247–P251.
22. Mulley G (1974). Proguanil and mouth ulcers. *Lancet*, **2**, 873.
23. Daniels AM (1986). Mouth ulceration associated with proguanil. *Lancet*, **i**, 269.
24. Boots M, Phillips M, Curtis JR (1982). Megaloblastic anaemia and pancytopenia due to proguanil in patients with chronic renal failure. *Clin Nephrol*, **18**, 106–108.
25. Drysdale SF, Phillips-Howard PA, Behrens RH (1990). Proguanil, chloroquine, and mouth ulcers. *Lancet*, **355**, 164.
26. *International Travel and Health: vaccination requirements and health advice* (1994). (Geneva: World Health Organization).
27. Rombo L (1994). Goda skäl att sänka dosen av proguanil. *Läkartidningen*, **91**, 3246.

Pyrantel

Chemical structure

Pyrantel

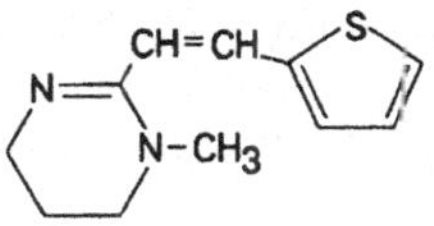

Physical properties

Pyrantel: MW 206. Embonate (also called pamoate): MW 595; pKa not known. Practically insoluble in water.

Pharmacology and mechanism of action

Pyrantel is a pyrimidine derivative with a broad spectrum anthelminthic activity.

Its mechanism of action is similar to that of bephenium and levamisole (see under Levamisole).

Pharmacokinetics

A specific analytical method has not been described.

The drug is poorly absorbed from the gastrointestinal tract with over half of the dose being recovered unchanged in the faeces. Less than 4% of the administered dose is recovered unchanged or as metabolites in the urine (1).

Clinical trials

A single dose of 10 mg/kg cured 80–100% of patients infected with *Ascaris lumbricoides* (2–4), *Enterobius vermicularis* (5) or *Ancylostoma duodenale* (2, 3, 6).

In patients infected with *Necator americanus*, a cure rate of 80% was achieved with a dose of 20 mg/kg daily for three days (7), but the efficacy was lower using a single dose (8).

Indications

Infections with *Ascaris lumbricoides*, *Enterobius vermicularis*, *Ancylostoma duodenale* and *Necator americanus*.

Pregnancy and lactation

Teratogenicity has not been reported in rabbits and mice (9, 10). Documentation in man is lacking. Pyrantel is probably safe during pregnancy, but its use should be postponed until after the first trimester.

Its excretion into breast milk is unknown, but it is unlikely that clinically significant concentrations would reach the milk since the drug is poorly absorbed.

Side effects

Pyrantel is well tolerated in children and adults. Clinical studies carried out in several regions of the world have demonstrated mild and transient side effects.

In a literature survey (11) of 1506 patients treated with 10 mg/kg of the drug, 299 subjects (20%) experienced side effects. The side effects were largely mild and transient. Frequent side effects were: nausea (4%), diarrhoea (4%), abdominal pain (4%), headache (3%) and vomiting (2%). Rare side effects included anorexia, fever, drowsiness, insomnia, skin rashes and elevation of serum aspartate aminotransferase (SGOT).

Contraindications and precautions

There are no known contraindications to the drug.

Interactions

Pyrantel antagonizes the effects of piperazine *in vitro* (12), and potentiates the actions of levamisole in pigs (13). The clinical significance of these interactions is as yet unknown.

Dosage

Infections with Ascaris lumbricoides, Ancylostoma duodenale and Enterobius vermicularis
Adults and children
10 mg base/kg as a single dose.

Infections with Necator americanus

Adults and children
20 mg base/kg as a single daily dose for 2 days.

Preparations

Available as pyrantel embonate: 725 mg embonate equals 250 mg base.

- Antiminth® (Pfizer). Oral suspension 50 mg base per ml.
- Combantrin® (Pfizer). Oral suspension 50 mg base per ml. Tablets (chewable) 250 mg base. Tablets (scored) 125, 250 mg base.
- Helmintox® (Innothera). Tablets 125, 250 mg base. Oral suspension 50 mg per ml.

References

1. Kimura Y, Kume M (1971). Absorption, distribution, excretion and metabolism of pyrantel pamoate. *Pharmacometrics*, **5**, 347–358.
2. Farahmandian I, Arfaa F, Jalali H, Reza M (1977). Comparative studies on the evaluation of the effect of new anthelminthics on various intestinal helminthiasis in Iran. *Chemother*, **23**, 98–105.
3. Islam N, Naseem A, Chowdhury A (1976). Mebendazole and pyrantel pamoate as broad-spectrum anthelminthics. *Southeast Asian J Trop Med Pub Health*, **7**, 81–84.
4. Sinniah B, Sinniah D (1981). The anthelminthic effects of pyrantel pamoate, oxantel-pyrantel pamoate, levamisole and mebendazole in the treatment of intestinal nematodes. *Ann Trop Med Parasitol*, **75**, 315–321.

5. Nassif S, Bell WJ, Prescott JEG (1974). Comparison of pyrantel pamoate syrup and pyrvinium pamoate syrup in the treatment of enterobiasis in Egypt. *J Trop Med Hyg*, **77**, 270–271.
6. Migasena S, Suntharasamai P, Harinasuta T (1978). Mebendazole, tetrachlorethylene and pyrantel pamoate in the treatment of hookworm infections. *Ann Trop Med Parasit*, **72**, 199–200.
7. Chege SW, Gitoho F, Wanene GSN, Mwega VJ, Rees PH, Kinyanjui H (1974). Single dose treatment of hookworm in Murang's District. *E Afr Med J*, **51**, 60–62.
8. Kale OO, Bammeke AO, Nwankwo EO (1982). Field trials of pyrantel pamoate (Combantrin) in *Ascaris*, hookworm, and *Trichuris* infections. *Afr J Med Sci*, **11**, 23–31.
9. Owaki Y, Sakai F, Momiyama H (1971). Teratogenic studies on pyrantel pamoate in rabbits. *Pharmacometrics*, **5**, 33–39.
10. Owaki Y, Sakai F, Momiyama H (1971). Teratogenic studies on pyrantel pamoate in rats. *Pharmacometrics*, **5**, 41–50.
11. Fossati C (1980). Sull'azione del pirantel-pamoato nella terapia delle infestationi da elminti. *Clin Ter*, **93**, 713–717.
12. Terada M, Fujiu Y, Sano M (1983). Studies on chemotherapy of parasitic helminths (XVII). Effects of pyrantel on the motility of various parasitic helminths and isolated host tissues. *Experientia*, **39**, 1020–1022.
13. Hsu W (1981). Drug interactions of levamisole with pyrantel tartrate and dichlorvos in pigs. *Am J Vet Res*, **42**, 1912–1914.

Pyrimethamine

Chemical structure

Pyrimethamine

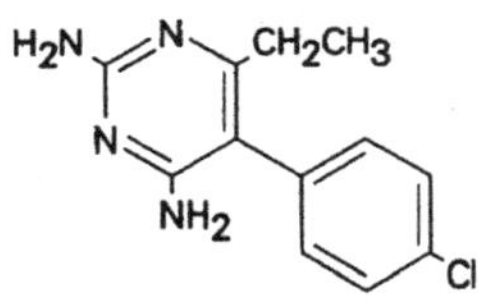

Physical properties

MW: 249; pKa: 7.3. The drug is practically insoluble in water. It should be stored in air-tight containers and be protected from light.

Pharmacology and mechanism of action

Pyrimethamine is a diaminopyrimidine which is structurally related to trimethoprim. It is effective against erythrocytic stage of *Plasmodium (P) falciparum* and less so against *P. vivax*, *P. ovale* and *P. malariae*. Pyrimethamine also inhibits the sporogony in the mosquito, resulting in a decrease of transmission of the infection within the community (1).

The mechanism of action of pyrimethamine is related to its inhibition of dihydrofolic reductase necessary for the folic acid synthesis in the parasite. Pyrimethamine acts slowly and is not recommended as monotherapy for acute malaria attacks. Resistance to pyrimethamine developed soon when the drug was used on a large scale as monoprophylaxis (1). In resistant strains, the enzyme dihydrofolic reductase binds to pyrimethamine several hundred times less than in sensitive strains (2). This high grade resistance is probably a one-step mutation and cannot be overcome by increasing the dose. However, when combined with long-acting sulphonomides (sulphadoxine), the effect of pyrimethamine is potentiated and the risk of developing resistant strains is far less.

Pharmacokinetics

Specific HPLC methods have been described (3–5).

Pyrimethamine in combination with sulphadoxine (Fansidar) is given orally as well as parenterally (i.m.). Absolute bioavailability of pyrimethamine is not known, but it is presumed to be completely absorbed. Peak plasma levels are usually reached within 4 hours of oral administration (6). After intramuscular injection, the drug is more slowly absorbed and peak plasma levels are reached after 1–2 days (7). Pyrimethamine has a volume of distribution of 2 l/kg and concentrations in plasma and whole blood are similar (6). More than 90% of the drug is bound to plasma proteins (8).

Pyrimethamine is metabolized in the liver but also excreted unchanged in the urine (9). The mean plasma elimination half-life is around 4 days (4, 7, 10).

Clinical trials

Pyrimethamine is used in combination with sulphadoxine (Fansidar) (cf. Sulphadoxine: Clinical trials). Pyrimethamine in combination with dapsone (Maloprim) has been implicated with a high risk of agranulocytosis and is not recommended for use any more (11).

Indications

Pyrimethamine together with sulphadoxine (Fansidar) is used in the treatment of *P. falciparum* malaria (cf. Sulphadoxine: Indications). Pyrimethamine is also valuable in the treatment of toxoplasmosis.

Pregnancy and lactation

Pyrimethamine is teratogenic in rats (12). Teratogenicity in man is not confirmed although there has been one suspected case of dysmorphogenesis where the mother had been treated with the drug during pregnancy (13). Pyrimethamine is used in combination with sulphadoxine and they are not contraindicated during pregnancy (cf. Sulphadoxine: Pregnancy and lactation).

Appreciable amounts of pyrimethamine is excreted into breast milk, but no adverse reactions have been reported in infants exposed to the drug (14).

Side effects

Pyrimethamine in combination with sulphadoxine (Fansidar) can cause severe cutaneous adverse reactions (cf. Sulphadoxine: Side effects). Agranulocytosis occurs quite frequently (1/2000) and fatalities have been reported when pyrimethamine is combined with dapsone (11). When given alone, life-threatening adverse reactions are very rare and the drug is generally well tolerated. Megaloblastic anaemia may, however, occur during long-term treatment with high doses (i.e. for toxoplasmosis) and can be prevented by folinic acid supplementation (9).

Contraindications and precautions

During long-term treatment with high doses, folinic acid supplement is usually given.

Dosage

For malaria treatment or prophylaxis pyrimethamine should be combined with sulphadoxine (cf. Sulphadoxine: Dosage).

Preparations

Pyrimethamine combined with sulphadoxine.

- Fansidar® (Roche). Tablets (pyrimethamine 25 mg plus sulphadoxine 500 mg). Solution for intramuscular injection (pyrimethamine 10 mg/ml and sulphadoxine 200 mg/ml).

References

1. Black RH, Canfield CJ, Clyde DF, Peters W, Wernsdorfer WH (1986). *Chemotherapy of Malaria*, 2nd edn, edited by L. J. Bruce-Chwatt. (Geneva: World Health Organization), pp. 77–80.

2. *The biology of malaria parasites*. Technical Report Series no 743 (1987). (Geneva: World Health Organization).
3. Bergqvist Y, Eriksson M (1985). Simultaneous determination of pyrimethamine and sulphadoxine in human plasma by high-performance liquid chromatography. *Trans R Soc Trop Med Hyg*, **79**(3), 297–301.
4. Midskov C (1984). High-performance liquid chromatographic assay of pyrimethamine, sulfadoxine and its N4-acetyl metabolite in serum and urine after ingestion of Suldox. *J Chromatogr*, **308**, 217–227.
5. Edstein MD, Lika ID, Chongsuphajaisiddhi T, Sabchareon A, Webster HK (1991). Quantitation of Fansimef components (mefloquine + sulfadoxine + pyrimethamine) in human plasma by two high-performance liquid chromatographic methods. *Ther Drug Monit*, **13**(2), 146–151.
6. Edstein MD (1987). Pharmacokinetics of sulfadoxine and pyrimethamine after Fansidar administration in man. *Chemotherapy*, **33**(4): 229–233.
7. Winstanley PA, Watkins WM, Newton CRJC, Nevill C, Emberu E, War PA, Waruiru CM, Mwangi IN, Warrell DA, Marsh K (1992). The disposition of oral and intramuscular pyrimethamine/sulphadoxine in Kenyan children with high parasitemia but clinically non-severe *falciparum* malaria. *Br J Clin Pharmacol*, **33**, 143–148.
8. Rudy AC, Poynor WJ (1990). Binding of pyrimethamine to human plasma proteins and erythrocytes. *Pharm Res*, **7**(10), 1055–1060.
9. Pyrimethamine. *Therapeutic Drugs*, edited by Sir Colin Dollery (1991), (London: Churchill Livingstone), pp. P314–P317.
10. Hellgren U, Angel VH, Berqvist Y, Arvidsson A, Segundo J, Forero-Gomez J, Rombo L (1990). Plasma concentrations of sulphadoxine-pyrimethamine and of mefloquine during regular long term malaria prophylaxis. *Trans R Soc Trop Med Hyg*, **84**, 46–49.
11. Friman G, Nyström-Rosander C, Jonsell G, Björkman A, Svendsrup B (1985). Agranulocytosis associated with malaria prophylaxis with Maloprim. *BMJ*, **286**, 1244–1245.
12. Coleman RD (1974). The incidence and development of cleft palate in rats following the maternal ingestion of pyrimethamine. *Anat Rec*, **178**, 332–333.
13. Harpy JP, Darbois Y, Lefèbvre G (1983). Teratogenicity of pyrimethamine. *Lancet*, **ii**, 399.
14. Clyde DF, Shute GT, Press J (1956). Transfer of pyrimethamine in human milk. *J Trop Med Hyg*, **59**, 277–284.

Pyrvinium Pamoate
(Viprynium Pamoate)

Chemical structure

Pyrvinium Pamoate

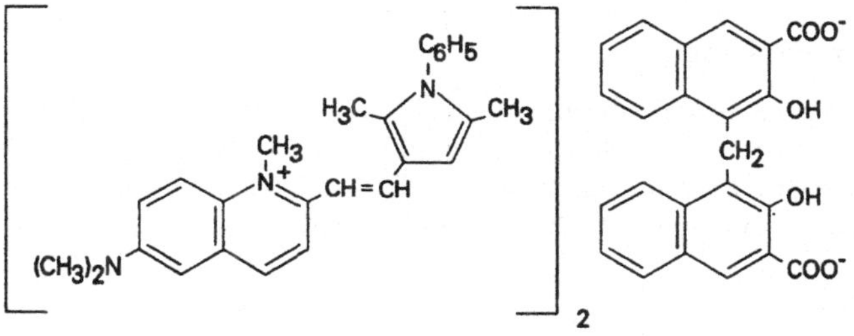

Physical properties

MW 384 (quaternary ammonium compound). Pamoate (also called embonate): MW 1151. The drug is practically insoluble in water. It should be stored in air-tight containers and be protected from light.

Pharmacology and mechanism of action

Pyrvinium is a quaternary ammonium derivative of a cyanine dye with marked effect against *Enterobius vermicularis*.

The mechanism of action is not well known. The drug inhibits oxidative metabolism in the worms and interferes with the absorption of glucose in intestinal helminths. Since most intestinal worms are dependent on anaerobic carbohydrate metabolism, this may be an important action of the drug (1).

Pharmacokinetics

A specific analytical method is not available.

Pyrvinium is apparently not absorbed in man. In one study (2), pyrvinium pamoate, tablets and suspension, was administered as single 350 mg doses to 12 healthy male volunteers in a cross-over design to determine whether there had been any systemic absorption. Up to 4 days after administration, there was no evidence of the drug in blood and urine as determined by spectrofluorometric assay (2). Metabolic studies in rats showed minute quantities of drug in the liver and plasma but not any of the metabolites (2).

Clinical trials

In a few open studies, pyrvinium in a single dose of 5 mg/kg base cured 90–100% of infections with *Enterobius vermicularis* (3, 4). The drug does not seem to have a therapeutic effect against other helminths.

Indications

Infections caused by *Enterobius vermicularis*. Pyrvinium is largely replaced today by more effective and safer drugs such as albendazole and mebendazole.

Pregnancy and lactation

Pyrvinium is probably safe during pregnancy, however, its use should be postponed until after the first trimester since it can aggravate nausea and vomiting during this period. Its excretion into breast milk is unknown, but it is unlikely that significant concentrations would reach the milk since the drug is poorly absorbed.

Side effects

Gastrointestinal disturbances such as nausea, vomiting, dyspepsia, indigestion and abdominal pain are frequent. Allergic reactions and photosensitivity have been reported. Pyrvinium stains the stool bright red and may stain clothing if vomiting occurs. Patients should be informed about this.

Contraindications

There are no known contraindications to the drug.

Dosage

5 mg/kg pyrvinium as a single dose. Treatment should be repeated after 2–3 weeks. It is wise to treat the whole family.

Preparations

Pyrvinium pamoate (embonate): 450 mg of the pamoate is equivalent to 300 mg of pyrvinium.

- Povanyl® (Parke-Davis). Oral solution 10 mg pyrvinium per ml. Tablets 50 mg pyrvinium.
- Vanquin® (Parke-Davis). Oral solution 10 mg pyrvinium per ml. Tab pyrvinium.

References

1. Buchanan RA, Barrow WB, Heffelfinger JC, Kinkel AW, Smith TC, Turner JL (1974). Pyrvinium pamoate. *Clin Pharmacol Ther*, **16**, 716–719.
2. Smith TC, Kinkel AW, Gryczko CM, Goulet JR (1976). Absorption of pyrvinium pamoate. *Clin Pharmacol Ther*, **19**, 802–806.
3. Bumbalo TS, Plummer LJ, Warner JR (1960). A clinical evaluation of four oxyuricides. *Am J Dis Child*, **99**, 617–621.
4. Royer A, Berdnikoff K (1962). Pinworm infestation in children: The problem and its treatment. *Can Med Assoc J*, **86**, 60–65.

Quinine

Chemical structure

Quinine

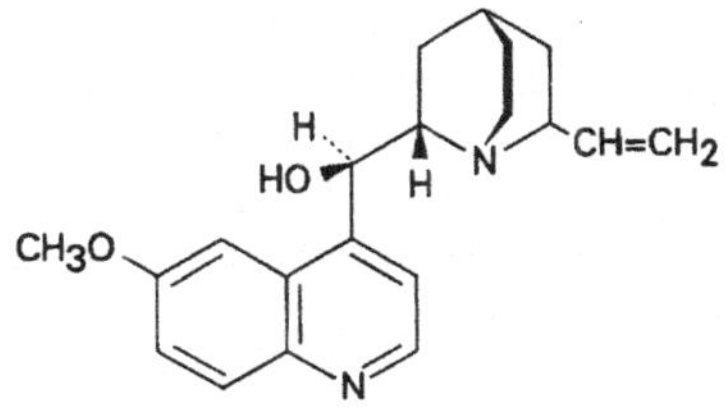

Physical properties

Base: MW 324; hydrochloride (dihydrate): MW 397; dihydrochloride: MW 397; bisulphate (heptahydrate): MW 549; sulphate (dihydrate): MW 783; formate: MW 380; pKa 4.1, 8.5. The quinine salts are moderately soluble in water with the exception of the sulphate which is only slightly soluble. The drug should be stored in air-tight containers and be protected from light.

Pharmacology and mechanism of action

Quinine is the principal alkaloid of cinchona bark. The cinchona bark was first used against fever in Peru, probably around 1630, but the compound may have been used much earlier by the native Indians. Soon thereafter it was introduced into Europe (1). Quinine is a stereoisomer of quinidine, which has similar antimalarial properties. It is a potent schizontocidal agent against all human plasmodial species. It is also gametocytocidal against *P. vivax*, *P. ovale*, and *P. malariae* but not against *P. falciparum* (1).

The mechanism of action is probably, as for chloroquine, an inhibition of haem polymerase (cf. Chloroquine) (2).

Pharmacokinetics

Spectrophotofluorometric methods have been widely used (3, 4) but they are unspecific as they co-determine quinine metabolites. Several specific HPLC methods have been described (5, 6, 7).

Quinine is given orally and parenterally. The oral bioavailability is high (76–88%) in healthy volunteers and in patients with uncomplicated malaria (8, 9, 10) irrespective of whether the hydrochloride, sulphate or ethylcarbonate salts are given (11). After oral administration, peak plasma levels are reached after 1 to 3 hours (9, 10, 11).

In healthy volunteers, the apparent volume of distribution is approximately 2 l/kg and the systemic clearance 150 ml/min (12).

Quinine is a weak base which is mainly bound to α_1-acid glycoprotein in plasma. The binding capacity in plasma is concentration dependent but also depends on the concentration of α_1-acid glycoprotein which makes comparisons difficult between studies (13). One month following recovery from malaria the protein binding was 89% (14).

The drug passes the blood–brain barrier, and in cerebral malaria the CSF concentrations were 4–7% of those in plasma (14, 15). Quinine is mainly eliminated by hydroxylation with a plasma elimination half-life of 7–11 hours (16, 17). Less than 20% of the total dose administered is recovered in the urine as the parent drug (18).

The pharmacokinetics of quinine is significantly altered during malaria infection. The plasma concentration of the acute phase reactant α_1-acid glycoprotein are increased in severely ill malaria patients compared to uncomplicated infections, and three times higher than in healthy individuals (19). As a consequence, the protein binding is increased. Total plasma concentrations of quinine in healthy volunteers are approximately 50% higher than in convalescence, but the free unbound concentration remains relatively unchanged (20). The apparent volume of distribution and the total systemic clearance are reduced (20) and parallel the disease severity (15). Half-lives decreased from 17 to 11 hours at convalescence in patients with uncomplicated *falciparum* malaria (15). Oral bioavailability is probably not significantly reduced (20). The increased plasma protein binding alone can cause these pharmacokinetic changes, but one study has also demonstrated a reduction in free quinine clearance (20).

Clinical trials

Quinine and the other cinchona alkaloids quinidine, cinchonine and cinchonidine are all effective against malaria. The first clinical trial consisting of 3617 patients was conducted in 1866–1868 in which the efficacy of the four compounds was compared. It was reported that the effect of all four compounds was roughly equivalent giving a clinical cure of above 98% (21). The predominant use of quinine after about 1890 was due to a change from South American to Javan cinchona bark which contained a higher proportion of quinine rather than a consideration of a greater efficacy (22).

When chloroquine became available in the late 1940s it replaced quinine all over the world, including Africa, despite the fact that quinine was still effective. With the spread of chloroquine-resistant *falciparum* malaria to almost all endemic countries the use of quinine has returned again.

In Thailand, the cure rate with quinine 600 mg 3 times daily for 7 days was reduced from 100% to 70% between 1963 and 1980 (23). The drug was then combined with tetracycline and the cure rate increased to over 90%. Despite the combination with tetracycline (4 mg/kg 4 times daily), the clinical and parasitological response decreased in severe *falciparum* malaria in Thailand from 1981 to 1992. The mortality, however, was unchanged, and it was concluded that quinine still remains an effective treatment although this will probably change as resistance increases (24).

In Africa, the antimalarial immunity in the indigenous population is stronger than elsewhere in the world, and there are only sporadic reports about reduced sensitivity to quinine *in vitro*. Lower doses and/or reduced treatment times have been used but not systematically evaluated. In patients with uncomplicated *falciparum* malaria, 10 mg/kg 3 times daily for only 3 days was effective in Zaire (25). In Madagascar, a crude quinine extract (Quinimax) was effective in an oral dose of 10 mg/kg 3 times daily for 3 days (26).

In Kenyan children with severe malaria, a low i.v. dose of quinine (10 mg/kg loading, then 5 mg every 12 hours) was less effective than standard treatment (20 mg/kg loading, then 10 mg every 12 hours) given i.v. or as intramuscular injections. When the children improved, oral treatment was given. The total treatment time was 5 days (27). The authors concluded that a reduction in dose is not appropriate, but intramuscular quinine can be used when the i.v. administration is not possible. In Nigerian children with cerebral malaria, the standard 7 day regimen (10 mg/kg 3 times daily) was effective (28).

Indications

Quinine is the drug of choice in the treatment of severe and complicated chloroquine-resistant *P. falciparum* malaria. It is also useful for the treatment of non-severe chloroquine-resistant cases.

Pregnancy and lactation

Quinine is not teratogenic in man with the recommended doses against malaria. The drug has been given to induce abortion but there is no evidence of an oxytocic effect in late pregnancy (29). Quinine is regarded as the drug of choice in the treatment of complicated *P. falciparum* malaria during pregnancy, but regular monitoring of the blood glucose is warranted.

Quinine is excreted into breast milk attaining concentrations of about one-fourth of that in plasma. The amount taken up by the infant is below the threshold to affect the parasite but may suffice to cause hypersensitivity reactions (30).

Side effects

The side effects of quinine commonly seen at therapeutic concentrations are known as cinchonism. In its mild form they include ringing in the ears (tinnitus), slight impairment of hearing, headache and nausea. The impairment of hearing is concentration-dependent and reversible (31). More severe manifestations are vertigo, vomiting, abdominal pain, diarrhoea, marked auditory loss and different visual symptoms like diplopia and changed colour perception but also loss of vision. The visual disturbances are probably caused by ischemia in the retina and the optic nerve, and this can cause optic atrophy. In acute intoxication, CNS symptoms such as excitement, confusion, delirium, and hyperpnoea may occur, and permanent visual and hearing deficits are not uncommon. Quinine may aggravate hypoglycaemia due to malaria. Less frequent but more serious side effects of quinine include skin manifestations, asthma, thrombocytopenia, haemolysis, hepatic injury and psychosis (32, 33). Patients with severe malaria attain and tolerate higher concentrations due to the concomitant reduction in free fraction.

Contraindications and precautions

Quinine should be avoided in patients who are hypersensitive to the drug and should not be given to patients with optic neuritis and those with myasthenia gravis since it can aggravate these conditions. Digoxin clearance is decreased by quinine and the two drugs should not be combined unless plasma concentration monitoring of digoxin is feasible. Quinine causes ECG changes after large doses, and patients with cardiac diseases must be treated with caution. There is a possible risk for increased cardiovascular toxicity when quinine is given to patients taking mefloquine prophylaxis or to those who have received mefloquine treatment within the last two weeks, and continuous cardiovascular monitoring is recommended (33). Diabetic patients may need special monitoring. Dosage adjustments may be needed in patients with liver diseases (34) and older subjects (35).

Interactions

Quinine shares most of the actions of quinidine, and most of the drug interactions seen with quinidine may be encountered with quinine as well. Quinine increases digoxin plasma levels, probably by reducing its non-renal clearance. Cimetidine has been reported to reduce the clearance of quinine and prolong its elimination half-life (32).

Dosage

Quinine is usually available as hydrocholoride, dihydrocholoride or sulphate and these are referred to below as 'quinine salts'. Each 10 mg salt contains approximately 8 mg base of quinine. However, if quinine bisulphate is used the total amount of salt has to be increased accordingly since 10 mg salt contains only 6 mg base of quinine.

A: Mild to moderately severe P. falciparum infections

1. In semi-immune individuals
Adults
600 mg oral quinine salt (500 mg base) or 10 mg/kg (8 mg base) every 12 hours for 5 days.
Children
10 mg/kg oral quinine salt (8 mg base) every 12 hours for 5 days.

2. In non-immune individuals
Adults
600 mg oral quinine salt (500 mg base) or 10 mg/kg (8 mg base) every 8 hours for 7 days.
Children
10 mg/kg oral quinine salt (8 mg base) every 8 hours for 7 days.

B: Severe P. falciparum infections

For intravenous infusion the required dose is preferably diluted in 5% glucose solution to avoid hypoglycaemia (if not available, physiological saline may be used) and is given in a large vein in a total volume of approximately 10 ml/kg (36).

Quinine can be administered by deep intramuscular injection if intravenous infusion is not possible. More diluted solutions, e.g. 60 mg/ml adjusted to neutral pH, are less painful

1. In semi-immune individuals
Children and adults
A loading dose of quinine salt 20 mg/kg (16 mg base/kg) intravenously over 4 hours. Thereafter quinine salt 10 mg/kg (8 mg base/kg) over 2–4 hours every 12 hours for 5–7 days until oral therapy is started.

2. In non-immune individuals
Children and adults
A loading dose of quinine salt 20 mg/kg (16 mg base/kg) intravenously over 4 hours. Thereafter quinine salt 10 mg/kg (8 mg base/kg) over 2–4 hours every 8 hours for 7–10 days until oral therapy is started.

In parts of Southeast Asia with reduced susceptibility for quinine it should be combined with tetracycline or doxycycline (not in children <8 years or pregnant women) as soon as the patient can take oral medication. An alternative is to give mefloquine 15 mg/kg not earlier than 8 hours after the last quinine dose or sulphadoxine pyrimethamine in standard treatment dose if parasites are susceptible to this combination.

Preparations

Numerous preparations (tablets, solution for injection) containing various quinine salts are available.

- Quinine hydrochloride (dihydrate). 123 mg equals 100 mg base.
- Quinine dihydrochloride. 123 mg equals 100 mg base.
- Quinine bisulphate (heptahydrate). 169 mg equals 100 mg base.
- Quinine sulphate (dihydrate). 121 mg equals 100 mg base.

References:

1. Black RH, Canfield CJ, Clyde DF, Peters W, Wernsdorfer WH (1986). Quinine. In: *Chemotherapy of Malaria*, 2nd edn, edited by L. Bruce-Chwatt (Geneva: World Health Organization).
2. Slater AFG, Cerami A (1992). Inhibition by chloroquine of a novel haem polymerase enzyme activity in malaria trophozoites. *Nature*, **355**, 167–169.
3. Brodie BB, Udenfriend S (1943). The estimation of quinine in human plasma with a note on the estimation of quinidine. *J Pharmacol Exp Ther*, **78**, 154–158.
4. Cramer G, Isaksson B (1963). Quantitative determination of quinidine in plasma. *Scand J Clin Lab Invest*, **15**, 553–556.
5. Edstein M, Stace J, Shann F (1983). Quantification of quinine in human serum by high-performance liquid chromatography. *J Chromatogr*, **278**, 445–451.
6. Rauch K, Ray J (1988). Improved high-performance liquid chromatographic method for the determination of quinine in plasma. *J Chromatogr*, **430**, 170–174.
7. Ericsson Ö, Fridén M, Hellgren U, Gustafsson LL (1993). Reversed-phase high-performance liquid chromatography determination of quinine in plasma, whole blood, and samples dried on filter paper. *Ther Drug Monit*, **15**, 334–447.
8. Sabcharoen A, Chongsuphajaisiddhi T, Attanath P (1982). Serum quinine concentrations following the initial dose in children with *falciparum* malaria. *Southeast Asian J Trop Med Pub Health*, **13**, 556–562.
9. Salako LA, Sowunmi A (1992). Disposition of quinine in plasma, red blood cells and saliva after oral and intravenous administration to healthy adult Africans. *Eur J clin Pharmac*, **42**, 171–174.
10. Paintaud G, Alván G, Ericsson Ö (1993). The reproducibility of quinine bioavailability. *Br J Clin Pharmacol*, **35**, 305–307.
11. Jamaludin A, Mohamad M, Navaratnam V, Selliah K, Tan SC, Wernsdorfer WH, Yuen KH (1988). Relative bioavailability of the hydrochloride, sulphate and ethyl carbonate salts of quinine. *Br J Clin Pharmacol*, **25**, 261–263.
12. White NJ (1988). Drug treatment and prevention of malaria. *Eur J Clin Pharmacol*, **34**, 1–14.
13. Mihaly GW, Ching MS, Klejn MB, Paule J, Smallwood RA (1987). Differences in the binding of quinine and quinidine to plasma proteins. *Br J Clin Pharmacol*, **24**, 769–774.
14. Silamut K, White NJ, Looareesuwan S, Warrell DA (1985). Binding of quinine to plasma proteins in *falciparum* malaria. *Am J Trop Med Hyg*, **34**, 681–686.
15. White NJ, Looareesuwan S, Warrel DA, Warrel MJ, Bunnag D, Harinasuta T (1982). Quinine pharmacokinetics and toxicity in cerebral and uncomplicated *falciparum* malaria. *Am J Med*, **73**, 564–572.
16. Jamaludin A, Mohamed M, Navaratnam V, Mohamed N, Yeoh E, Wernsdorfer WH (1988). Single-dose comparative kinetics and bioavailability study of quinine hydrochloride, quinidine sulphate, and quinidine bisulphate sustained-release in healthy male volunteers. *Acta Leyden*, **57**, 39–46.
17. White NJ, Chanthavanich P, Krishna S, Bunch C, Silamut K (1983). Quinine disposition kinetics. *Br J Clin Pharmacol*, **16**, 399–404.

18. Trenholme GM, Williams RL, Rieckman KH, Frischer H, Carson P (1976). Quinine disposition during malaria and during induced fever. *Clin Pharmacol Ther*, **19**, 459–467.
19. Silamut K, Molunto P, Ho M, Davis TME, White NJ (1991). alfa$_1$-acid glycoprotein (orosomucoid) and plasma protein binding of quinine in *falciparum* malaria. *Br J Clin Pharmacol*, **32**, 311–315.
20. Supanaranond W, Davis TME, Pukrittayakamee S, Silamut K, Karbwang J, Molunto P, Chanond L, White NJ (1991). Disposition of oral quinine in acute *falciparum* malaria. *Eur J Clin Pharmacol*, **40**, 49–52.
21. Madras Cinchona Commission (1870) *Return East India (Cinchona cultivation)*, (London: H.M.'s Stationery Office).
22. Howard BF (1931). Some notes on the cinchona industry. *Chemical News*, **142**, 129–133.
23. Harinasuta T, Bunnag D (1984). Drug resistant malaria with special reference to chemotherapy. *Mosquito-Borne Diseases Bulletin*, **1**, 23–30.
24. Pukrittayakamee S, Supanaranond W, Looareesuwan S, Vanijanonta S, White NJ (1994). Quinine in severe *falciparum* malaria: evidence of declining efficacy in Thailand. *Trans R Soc Trop Med Hyg*, **88**, 324–327.
25. Greenberg AE, Ngueyn-Dinh P, Davach F, Yemvula B, Malanda N, Nzeza M, Williams SB, Zwart JF, Nzeza M (1989). Intravenous quinine therapy of hospitalized children with *Plasmodium falciparum* malaria in Kinshasa, Zaire. *Am J Trop Med Hyg*, **40**, 360–364.
26. Deloron P, Lepers JP, Andriamangatiana-Rason MD, Coulanges P (1990). Short-term oral cinchona alkaloids regimens for treatment of *falciparum* malaria in Madagascar. *Trans R Soc Trop Med Hyg*, **84**, 54.
27. Pasvol G, Newton CRJC, Winstanley PA, Watkins WM, Peshu NM, Were JBO, Marsh K, Warrell DA (1991). Quinine treatment of severe *falciparum* malaria in African children: a randomized comparison of three regimens. *Am J Trop Med Hyg*, **45**(6), 702–713.
28. Walker O, Salako LA, Omokhodion SI, Sowunmi A (1993). An open randomized comparative study of intramuscular artemether and intravenous quinine in cerebral malaria in children. *Trans R Soc Trop Med Hyg*, **87**, 564–566.
29. Looareesuwan S, White NJ, Karbwang J, Turner RC, Phillips RE, Kietinun S, Rackow C, Warrell DA (1985). Quinine and severe *falciparum* malaria in late pregnancy. *Lancet*, **2**, 4–8.
30. Looareesuwan S, White NJ, Kamolrat S, Phillips RE, Warrel DA (1987). Quinine and severe *falciparum* malaria in late pregnancy. *Acta Leyden*, **55**, 115–120.
31. Karlsson KK, Hellgren U, Alván G, Rombo L (1990). Audiometry as a possible indication of quinine plasma concentrations during treatment of malaria. *Trans R Soc Trop Med Hyg*, **84**, 765–767.
32. Antimalarials. Martindale, *The Extra Pharmacopoeia*, 30th edn (1993), (London: Pharmaceutical Press), pp. 408.
33. Quinine. *Therapeutic Drugs*, edited by Sir Colin Dollery (1991), (London: Churchill Livingstone), pp. Q8–Q13.
34. Karbwang J, Thanavibul A, Molunto P, Na Bangchang K (1993). The pharmacokinetics of quinine in patients with hepatitis. *Br J Clin Pharmacol*, **35**, 444–446.
35. Wanwimolruk S, Chalcroft S, Coville PF, Campbell AJ (1991). Pharmacokinetics of quinine in young and elderly subjects. *Trans R Soc Trop Med Hyg*, **85**, 714–717.
36. *WHO Model Prescribing Information. Drugs used in parasitic diseases* (1990), (Geneva: World Health Organization).

Sulphadoxine

Chemical structure

Sulphadoxine

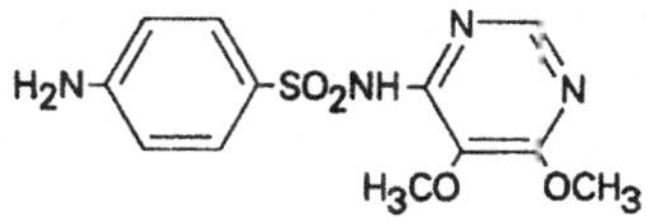

Physical properties

MW: 310; pKa: 6. The drug is only slightly soluble in water. It should be protected from light.

Pharmacology and mechanism of action

The efficacy of sulphadoxine for the treatment of human malaria was first reported in 1964 (1). Soon thereafter it was found that potentiation took place when sulphadoxine was combined with pyrimethamine for treatment of malaria and monotherapy was abandoned.

Malaria parasites synthesize their folate co-factors and cannot use dietary folic acid as the human host can. Sulphadoxine competes with para-aminobenzoic acid (PABA) for binding to the enzyme dihydropteroate synthetase in the synthesis of dihydropteroate which is an essential substance for the formation of folic acid (2). It is active against asexual blood forms of *P. falciparum* but less active against other species. The action is too slow to be used alone for malaria treatment (3).

Pharmacokinetics

The classical method to determine sulphadoxine is spectrophotometric (4), but this cannot separate the mother substance from non-acetylated metabolites. A simple and specific HPLC method for determination of sulphadoxine has been described (5). Methods based on HPLC analysis of capillary blood dried on filter paper are also available (6).

Sulphadoxine is given together with pyrimethamine (Fansidar) orally as well as parenterally (i.m.). The oral bioavailability is high (7). After oral administration, peak plasma levels are reached after 4–5 hours (7, 8). About 90% of the drug is bound to plasma proteins, predominantly to albumin (5). Penetration of sulphadoxine into erythrocytes is low in healthy volunteers, but it appears to be concentrated in malaria-infected erythrocytes (9, 10). The drug enters the CSF with a concentration of one-eighth to one-fifth of that in plasma (7). The volume of distribution is 0.12–0.15 l/kg (8, 11).

Only a small proportion of sulphadoxine is metabolized and acetylated or glucuronidated (less than 10%). The metabolites are excreted faster than the parent drug and show higher concentrations in the urine (7, 8). Excretion studies of radioactively labelled sulphadoxine in man showed that approximately 90% is excreted with the urine and 10% with the stools (7).

The drug itself is eliminated slowly from the body with a mean plasma elimination half-life between 7 and 9 days (7, 8, 12). Sulphadoxine has a high rate of tubular reabsorption which contributes to its long persistence in the body (7). It is not known whether there are slow and rapid acetylators of sulphadoxine, but in a study from Thailand there was no difference in sulphadoxine or acetylated metabolite concentrations between responders and non-responders (13).

Clinical trials

Only a few years after the introduction of the sulphadoxine/pyrimethamine combination as first line treatment of *falciparum* malaria, a decrease in susceptibility was noted in Thailand (14). From this initial focus resistance has spread, and high-degree resistance is now frequent in the whole of Southeast Asia.

There have been sporadic reports of resistance against sulphadoxine/pyrimethamine in visitors to East Africa (15, 16). Recent treatment studies in indigenous populations, however, have demonstrated a high efficacy throughout the continent (12, 17, 18) except for Rwanda (19). As high-grade resistance can develop rapidly it is necessary to monitor the susceptibility for the sulphadoxine/pyrimethamine combination regularly. In South America there is a widespread, mostly low-grade, resistance.

Indications

Sulphadoxine is used only in combination with pyrimethamine for the treatment of *falciparum* malaria. It should generally not be used for malaria prophylaxis except perhaps in long-term travellers who have previously tolerated the combination.

Pregnancy and lactation

Although sulphadoxine crosses the placenta (20), no teratogenicity has been reported in mice, rats and rabbits (7). There have been no reports of human teratogenicity despite the widespread use of the drug during pregnancy. Theoretically, sulphadoxine, like some other sulphonamides, could cause kernicterus in the new-born infant if used during the end of pregnancy, but this has not been reported. Sulfadoxine is used in combination with pyrimethamine and they are not contraindicated during pregnancy, but folate supplementation may be required.

Sulphadoxine is excreted into breast milk in concentrations less than one-third of that in the plasma of the mother (7). However, because of its severe cutaneous reactions, breast feeding should be avoided.

Side effects

Sulphadoxine is usually well tolerated. Vomiting, skin rashes, pruritus and haematological reactions such as haemolysis and leucopenia occur (21). Hypersensitivity pneumonitis is reported (22, 23). Cases of liver injury alone (hepatitis of hepatocellular, mixed hepatocellular, or aggressive type) or as part of a generalized allergic syndrome are well known (21, 23), and one case of fatal hepatic failure has also been reported (24).

Severe cutaneous adverse reactions (erythema multiforme, Stevens–Johnson syndrome, or toxic epidermal necrolysis) have been reported in persons taking prophylactic doses of sulphadoxine/pyrimethamine (23, 25). Sulphadoxine has been incriminated as the most probable cause of these reactions. They all occurred within 7 weeks after start of prophylaxis. The reported incidence was 1/5,000–8,000 users in USA and approximately 1/10,000 users in Sweden with fatality rates of 1/11,000–25,000 and 1/50,000, respectively. In Mozambique, when sulphadoxine was given alone in a single dose for cholera prophylaxis to 149,000 inhabitants, a total of 22 cases of Stevens–Johnson syndrome was seen with 3 deaths (26).

Contraindications and precautions

The drug or its combination should not be given to patients allergic to sulphonamides. It should not be used in persons with severe blood, kidney or liver diseases.

Interactions

Increased impairment of folic acid synthesis and consequent haematological adverse effects may occur if trimethoprim or its combination with sulphonamide is administered concurrently. Sulphadoxine potentiates the action of warfarin and thiopentone (27).

Dosage

For malaria treatment or prophylaxis sulphadoxine should be combined with pyrimethamine.

Prophylaxis: Sulphadoxine + pyrimethamine (Fansidar)

Adults
1 tablet once a week.
Children

<4 years (5–10 kg)	¼ tablet once a week
4–9 years (11–30 kg)	½ tablet once a week
10–14 years (31–45 kg)	¾ tablet once a week

Treatment: Sulphadoxine + pyrimethamine (Fansidar)

Adults
3 tablets or 7.5 ml ampoule (i.m.)
Children

<4 years (5–10 kg)	½ tablet	1.25 ml ampoule (i.m.)
4–6 years (11–20 kg)	1 tablet	2.25 ml ampoule (i.m.)
7–9 years (21–30 kg)	1½ tablets	3.75 ml ampoule (i.m.)
10–14 years (31–45 kg)	2 tablets	5 ml ampoule (i.m.)

Preparations

Sulphadoxine combined with pyrimethamine.

- Fansidar® (Roche). Tablets (sulphadoxine 500 mg + pyrimethamine 25 mg), solution for intramuscular injection (sulphadoxine 200 mg/ml + pyrimethamine 10 mg/ml).

References

1. Laing ABG (1964). Antimalarial effect of sulphorthodimethoxine (Fanasil). *BMJ*, **2**, 1439–1440.
2. *The biology of malaria parasites*. Technical Report Series no. 743 (1987). (Geneva: World Health Organization).

3. *Chemotherapy of Malaria*. Monograph series No. 27, 2nd edn, (1986), (Geneva: World Health Organization).
4. Bratton AC, Marshall EK (1939). A new coupling component for sulfanilamide determination. *J Biol Chem*, **128**, 537–550.
5. Edstein MD, Lika ID, Chongsuphajaisiddhi T, Sabchareon A, Webster HK (1991). Quantitation of Fansimef components (mefloquine + sulfadoxine + pyrimethamine) in human plasma by two high-performance liquid chromatographic methods. *Ther Drug Monit*, **13**, 146–151.
6. Bergqvist Y, Hjelm E, Rombo L (1987). Sulfadoxine assay using capillary blood samples dried on filter paper – suitable for monitoring of blood concentrations in the field. *Ther Drug Monit*, **9**, 203–207.
7. Böhni E, Fust B, Reider J, Schaerer K, Havas L (1969). Comparative toxicological, chemotherapeutic and pharmacokinetic studies with sulphormethoxine and other sulphonomides in animals and man. *Chemother*, **14**, 195–226.
8. Edstein MD (1987). Pharmacokinetics of sulfadoxine and pyrimethamine after Fansidar administration in man. *Chemotherapy*, **33**, 229–233.
9. Berneis K, Boguth W (1976). Distribution of sulfonomides and sulfonomide potentiators between red blood cells, proteins and aqueous phases of the blood of different species. *Chemotherapy*, **22**, 390–409.
10. Dieckmann A, Jung A (1986). Mechanism of sulfadoxine resistance in *Plasmodium falciparum*. *Molecular and Biochemical Parasitology*, **19**, 143–147.
11. Portwich F, Büttner H (1964). Zur Pharmakokinetik eines langwirkenden Sulfonamids (4-Sulfanilamido-5,6-dimethoxypyridimidin) bei gesunden Menschen. *Klin Wochenschr*, **42**, 740–744.
12. Hellgren U, Kihamia CM, Bergqvist Y, Lebbad M, Rombo L (1990). Standard and reduced doses of sulphadoxine-pyrimethamine for treatment of *Plasmodium falciparum* malaria in Tanzania with determination of drug concentrations and susceptibility *in vitro*. *Trans R Soc Trop Med Hyg*, **84**, 469–473.
13. Sarikabuthi B, Keschamrus N, Noėypatimanond S, Weidekamm E, Leimer R, Wernsdorfer W, Kölle EU (1988). Plasma concentrations of sulphadoxine in healthy and malaria infected Thai subjects. *Acta Tropica*, **45**, 217–224.
14. Segal HE, Chinvanthananod P, Laixuthai B, Pearlmann EJ, Hall AP, Phintuyothin P, Amporn NA, Castaneda BF (1975). Comparison of diaminodiphenyl-sulphone pyrimethamine and sulfadoxine-pyrimethamine combinations in the treatment of *falciparum* malaria in Thailand. *Trans R Soc Trop Med Hyg*, **69**, 139–142.
15. Timmermans PM, Hess U, Jones ME (1982). Pyrimethamine/sulfadoxine resistant *falciparum* malaria in East Africa. *Lancet*, **i**, 1181.
16. Schapira A, Bygbjerg C, Jepsen S, Flachs H, Weis Bentzon M (1986). The susceptibility of *Plasmodium falciparum* to sulfadoxine and pyrimethamine: correlation of *in vivo* and *in vitro* results. *Am J Trop Med Hyg*, **35**, 239–245.
17. Salako LA, Adio RA, Sowunmi A, Walker O (1990). Parenteral sulphadoxine-pyrimethamine (Fansidar®): an effective and safe but under-used method of anti-malarial treatment. *Trans R Soc Trop Med Hyg*, **84**, 641–643.
18. Bloland PB, Redd SC, Kazembe P, Tembenu R, Wirima JJ, Campbell CC (1990). Co-trimoxazole for childhood febrile illness in malaria-endemic regions. *Lancet*, **337**, 518–520.
19. Garcia-Vidal J, Ngirabega J, Soldevila M, Navarro R, Bada JL (1989). Evolution of resistance of *Plasmodium falciparum* to antimalarial drugs in Rwanda, 1985–1987. *Trans R Soc Trop Med Hyg*, **83**, 490.
20. Fay R, Marx-ChemLa C, Leroux B, Harika G, Dupouy D, Quereux C, Choisy H, Pinon JM, Wahl P (1990). Passage transplacentaire de l'association pyriméthamine-sulfadoxine lors du traitement anténatal de la toxoplasmose congénitale. *Presse Méd*, **19**(44), 2036.
21. Hoigné R, Malinverni R, Sonntag R (1992). Sulfonomides, other folic acid antagonists and miscellaneous antibacterial drugs. In: *Meyler's Side Effects of Drugs*, 12th edn, edited by M. N. G. Dukes (Amsterdam: Elsevier), pp. 715–722.
22. Svanbom M, Rombo L, Gustafsson L (1984). Unusual pulmonary reaction during short term prophylaxis with pyrimethamine-sulphadoxine (Fansidar). *BMJ*, **1**, 1876.
23. Hellgren U, Rombo L, Berg B, Carlson J, Wiholm B-E (1987). Adverse reactions to sulphadoxine-pyrimethamine in Swedish travellers: implications for prophylaxis. *BMJ*, **295**, 365–366.
24. Zitelli BJ, Alexander J, Taylor S (1987). Fatal hepatic necrosis due to pyrimethamine-sulphadoxine (Fansidar). *Ann Intern Med*, **106**, 393–395.

25. Miller KD, Lobel HO, Satriale RF, Kirutsky JN, Stern R, Campell CC (1986). Severe cutaneous reactions among American travellers using pyrimethamine-sulphadoxine (Fansidar) for malaria prophylaxis. *Am J Trop Hyg*, **35**, 451–458.
26. Hernberg A (1985). Stevens–Johnson syndrome after mass prophylaxis with sulphadoxine for cholera in Mozambique. *Lancet*, **2**, 1072–1073.
27. Sulfadoxine. *Therapeutic Drugs*, edited by Sir Colin Dollery (1991), (London: Churchill Livingstone), pp. S115–S119.

Suramin

Chemical structure

Suramin

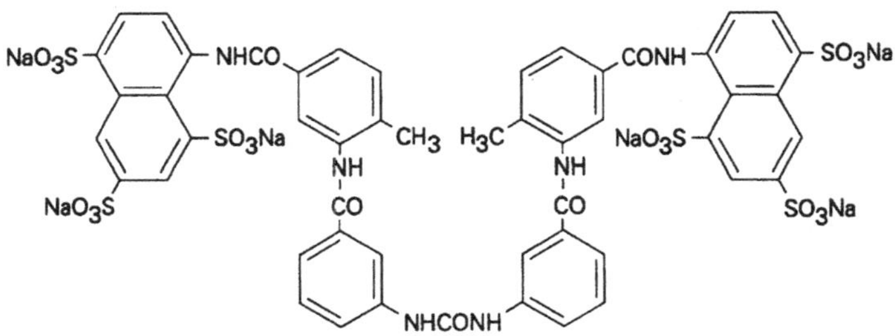

Physical properties

Acid: MW 1297; sodium salt: MW 1429; pKa not known. Freely soluble in water. Suramin deteriorates in air and should be injected immediately after preparation.

Pharmacology and mechanism of action

Suramin is a polysulphonated naphthylurea introduced in Germany in 1920 for the treatment of trypanosomiasis. The drug was later found to be an effective macrofilaricide in onchocerciasis. Today suramin is mainly used for the treatment of African trypanosomiasis. It is effective against early-stages of *Trypanosoma brucei gambiense* and *Trypanosoma brucei rhodesiense* without CNS involvement. However, in the case of *Trypanosoma brucei gambiense* pentamidine is generally preferred (1).

The mechanism of action of suramin is unknown. The drug has a broad spectrum of enzymatic actions, which is mainly due to its strong affinity for proteins. It interferes with the DNA–RNA replication mechanism of the cell, thus stopping cell growth. *In vitro*, suramin slowly inhibits the oxygen consumption of trypanosomes (1, 2). In *Brugia pahangi*, the drug acts on the surface of the intestinal epithelium resulting in ultrastructural changes (3). Suramin also impairs the *in vitro* infectivity of human immunodeficiency virus type I (HIV) (4), but the drug had little success in patients with AIDS and with different types of cancer (4, 5).

Pharmacokinetics

Specific HPLC methods have been described for the determination of suramin (6–9).

Suramin is poorly absorbed from the intestine and causes intense local irritation when given intramuscularly. It is thus always given by a slow intravenous injection. After a single weekly dose of 1 g for six weeks, plasma concentrations of 150–200 μg/ml were measured

(7). The drug has an apparent volume of distribution of around 54 l. About 99.7% is bound to plasma proteins (10). It is taken up by the reticuloendothelial cells and accumulates in the Kupfer cells of the liver as well as the epithelial cells of the proximal convoluted tubules of the kidney (1). Since suramin is a large polar compound it does not accumulate in red blood cells and its access into the CSF is insignificant (10). The plasma concentrations decline in a multiexponential fashion with a terminal plasma elimination half-life of around 60 days (10, 11). Total body clearance of the drug is low (0.5 ml/min) (12), of which about 80% is accounted for by renal clearance (10).

Clinical trials

No double-blind clinical trials have been reported. In an open study in Burkina Faso, 78 patients suffering from onchocerciasis were treated with suramin, 0.2 g initially followed by weekly doses of 0.4 g, 0.6 g, 0.8 g, and 1 g. Total doses varied between 3–4 g. After 3 months, the worm burden was reduced by about 90%. The drug was well tolerated (13).

Indications

Suramin is used in the treatment of early-stage infections of *Trypanosoma brucei gambiense* and *Trypanosoma brucei rhodesiense*. It is used prior to melarsoprol treatment to clear the blood and lymph of trypanosomes. Pentamidine is generally preferred for the treatment of early-stage *Trypanosoma brucei gambiense*.

Suramin is also the only drug available for effectively eliminating the adult filariae (macrofilariae) in onchocerciasis. It should only be used in individual cases.

Pregnancy and lactation

Teratogenicity has been reported in rats (14). Documentation is limited in man. A single woman in her 20th week of pregnancy treated with suramin has been reported. She gave birth to an apparently normal child (15). Because of the severity of the disease, suramin should not be withheld from a pregnant woman suffering from trypanosomiasis.

Its excretion into breast milk is unknown.

Side effects

Suramin is a toxic drug, and adverse reactions can be serious especially in malnourished patients (1). They are also more frequently observed in patients with onchocerciasis than in patients with trypanosomiasis. Adverse reactions due to suramin can be classified into three types:

1. Immediate reactions: with a frequency of about 0.1–0.3%. These include nausea, vomiting and loss of consciousness. Slight fever, acute urticaria and colic pain are other acute reactions. These reactions can be avoided by a slow i.v. injection. It is a general clinical practice to start with a small test dose to assess the patients tolerance.
2. Late reactions (3–48 hours after i.v. injection): these include fever up to 40 °C, photophobia and lacrimation. Flatulence, constipation and hyperaesthesia in the palms of the hands and the soles of the feet which can last for weeks.
3. Delayed reactions: these include kidney damage, which can occur several days after treatment and is a result of drug accumulation in the epithelial cells of the proximal convoluted tubules. Other delayed toxic reactions include dermatitis, stomatitis of the

exfoliative type, agranulocytosis, haemolytic anaemia, jaundice and severe diarrhoea which may be fatal.

During onchocerciasis treatment, reactions such as urticaria, itching, formation of deep abscesses and muscle pains are common which are due to the dying worm. Nodulectomy prior to suramin therapy decreases the severity of this reaction. These reactions should be treated with 1 mg of betamethasone (equivalent to 6 mg of prednisolone) 3 times daily for 3 days (1).

Contraindications and precautions

Suramin is a toxic drug and it should always be administered under medical supervision. Great caution should be exercised in malnourished patients and those with kidney diseases. Therapy should be discontinued or postponed in patients with heavy albuminuria with casts (1).

Interactions

There have been no reports.

Dosage and administration (16)

Several different dosage regimens have been reported, but any advantage of one over the other has not been shown. None seem to fit the pharmacokinetic profile of the drug. The drug should be administered by slow intravenous injection of a 10% aqueous solution. The first injection should be given with particular caution since a very few individuals may get severe idiosyncratic reactions, particularly during treatment of onchocerciasis.

Trypanosomiasis

Adults and children (mg/kg)

	Treatment days						
Treatment	1	3	5	11	17	23	30
A (radical cure)	5	10	20	20	20	20	20
B (before melarsoprol treatment)	5	10	20				

Onchocerciasis

Suramin is usually preceded by a course of diethylcarbamazine. Subsequently, a further course of diethylcarbamazine follows (see also under Diethylcarbamazine: Dosage). For adults, a total of 66.7 mg/kg should be administered in six incremental weekly doses apportioned as shown in the following table:

Week	1	2	3	4	5	6
Dose (mg/kg)	3.3	6.7	10.0	13.3	16.7	16.7

Preparations

- Suramin® (Bayer). Substance for injection 1 g.
- Germanin® (Bayer). Substance for injection 1 g.

References

1. Hawking, F (1978). Suramin: with special reference to onchocerciasis. *Adv Pharmacol Ther*, **15**, 289–322.
2. Fairlamb AH, Bowman IBR (1980). Uptake of the trypanocidal drug suramin by blood stream forms of *Trypanosoma brucei* and its effect on respiration and growth rate *in vivo*. *Mol Biochem Parasitol*, **1**, 315–333.
3. Howells RE, Mendis AM, Bray PG (1983). The mode of action of suramin on filarial worm *Brugia pahangi*. *Parasitology*, **87**, 29–48.
4. Broder S, Yarchoan R, Collins JM, Lane HC, Markham PD, Klecher RW, Redfield RR, Mitsuya H, Hoth DF, German E (1985). Effects of suramin on HTLV–III/LAV infection presenting as Kaposi's sarcoma or AIDS-related complex: clinical pharmacology and suppression of virus replication *in vivo*. *Lancet*, **ii**, 627–630.
5. Stein CA, LaRocca RV, Thomas R, McAtee N, Myers CE (1989). Suramin: an anticancer drug with a unique mechanism of action. *J Clin Oncol*, **7**, 499–508.
6. Edwards G, Rodick CL, Ward SA, Awadzi K, Orme ML'E, Breckenridge AM (1985). Determination of suramin in plasma by high-performance liquid chromatography. *J Chromatogr*, **343**, 224–228.
7. Klecker RW, Collins JM (1985). Quantification of suramin by reverse-phase ion pairing high performance liquid chromatography. *J Liq Chromatogr*, **8**, 1685–1696.
8. Ruprecht RM, Lorsch J, Trites DH (1986). Analysis of suramin plasma levels by ion-pair high-performance liquid chromatography under isocratic conditions. *J Chromatogr*, **378**, 498–502.
9. Tjaden UR, Reeuwijk HJEM, Van Der Greef J, Pattyn G, de Bruijn EA, Van Oosterom AT (1990). Bioanalysis of suramin in human plasma by ion-pair high-performance liquid chromatography. *J Chromatogr*, **525**, 141–149.
10. Collins JM, Klecker RW, Yarchoan R, Lane HC, Fauci AS (1986). Clinical pharmacokinetics of suramin in patients with HTLV III/LAV infection. *J Clin Pharmacol*, **26**, 22–26.
11. Van Boxtel CJ, Schattenkerk EFK, Van Den Berg M, de Graaf YP, AM, Danner SA (1986). Therapeutic monitoring of suramin in patients with AIDS. *Acta Pharmacol Toxicol*, **59**, 222.
12. Edwards, G Rodick, CL Ward SA, Awadzi K, Orme ML'E, Breckenridge AM (1986). Disposition of suramin in patients with onchocerciasis. *Acta Pharmacol Toxicol*, **59**, 222.
13. Rougemont A, Hien M, Thylefors B, Prost A, Schultz-key H, Rolland A (1984). Traitement de l'onchocercose par la suramine à faibles doses progressives dans les collectivités hyperendémiques d'Afrique occidentale. II: Résultats cliniques, parasitologiques et ophtalmologiques en zone de transmission interrompue. *Bull WHO*, **62**, 261–269.
14. Feeman JJ, Llyod JB (1986). Evidence that suramin and auro-thiomalate are teratogenic in rats by disturbing yolk sac-mediated embryonic protein nutrition. *Chem Biol Interact*, **58**, 149–160.
15. Lowenthal MN (1971). Trypanosomiasis successfully treated with suramin in a pregnant woman. *Med J Zambia*, **5**, 175–178.
16. *WHO Model Prescribing Information. Drugs used in parasitic diseases* (1990), (Geneva: World Health Organization).

Tetracyclines

Chemical structure

Tetracycline

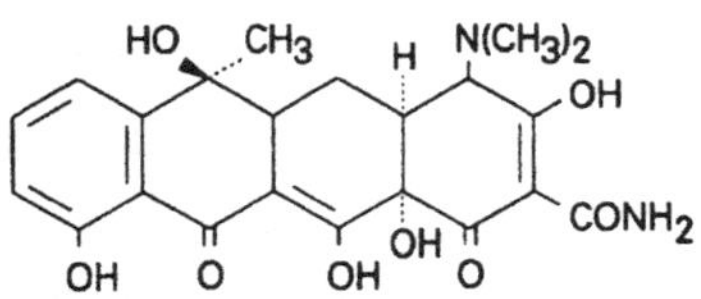

Doxycycline

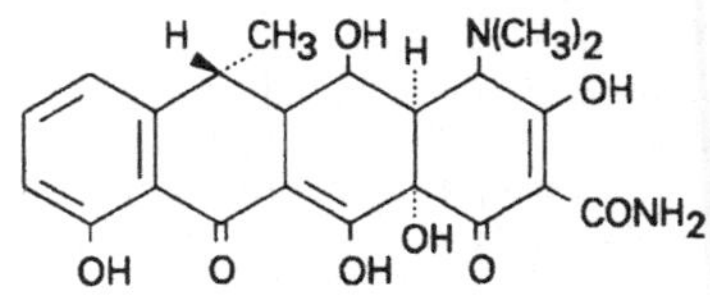

Physical properties

Tetracycline

Base: MW 444; hydrochloride: MW 481; pKa 3.3, 7.7 (acidic), 9.7 (basic). 1 g of the hydrochloride dissolves in 10 ml of water. Store in airtight containers. The drug should be protected from light.

Doxycycline

Base: Monohydrate MW 462; pKa 3.5, 7.7 (acidic), 9.5 (basic). Slightly soluble in water. Store in air-tight containers. The drug should be protected from light.

Pharmacology and mechanism of action

Tetracyclines are bacteriostatic antibiotics with broad-spectrum activity. They are primarily used for the treatment of chlamydia, rickettsia, mycoplasma, spirochete as well as infections due to Gram-positive and Gram-negative bacteria. For the treatment of malaria, tetracyclines have a potent but slow blood schizontocidal effect, thus they are used together with quinine. They are also effective against the primary tissue stages of *Plasmodium falciparum* but lack gametocytocidal effect (1).

The mechanism of action of tetracyclines in bacteria is inhibition of protein synthesis by binding to the S–30 ribosome, inhibiting the access of tRNA to the mRNA-ribosome complex (2). The mechanism of action against malaria is not known.

Pharmacokinetics

Specific HPLC methods are available for the determination of tetracycline (3) and doxycycline (4–5).

The oral bioavailability for tetracycline varies between 60% and 80%. Milk and milk products as well as antacids containing calcium, aluminium or magnesium impair the absorption of tetracycline by forming insoluble complexes (2, 6). Tetracycline should therefore be taken on an empty stomach. After a single oral dose of 250 mg, peak plasma levels of around 2 μg/ml are reached in 2–4 hours. The drug concentrates in the bile attaining concentrations 5 to 10 times higher than in plasma. It is weakly bound to plasma proteins (about 50%). The apparent volume of distribution is 1–2 l/kg. Tetracycline has a plasma elimination half-life of around 9 hours. Excretion is mainly as unchanged drug through the kidneys and faeces. About half of the dose is eliminated by the kidneys by glomerular filtration (2, 6).

Doxycycline has a high oral bioavailability (95%) and can be given with meals. Protein-binding of doxycycline is higher (80–90%) and the half-life considerably longer (15–25 hours) than for tetracycline (7). Doxcycline is not metabolized to any significant degree. Approximately 30–40% is eliminated renally and the rest is probably excreted with the bile. It does not accumulate in patients with renal failure which is thought to be due to its increased excretion in faeces in such patients (7).

Clinical trials

Tetracycline in combination with quinine has been used for the treatment of *falciparum* malaria in Thailand (8–10). There is, however, little information about the recent efficacy of the combination. In Thai adult patients with uncomplicated *falciparum* malaria, doxycycline 200 mg daily together with a total dose of 1250 mg mefloquine was highly effective and well tolerated (11).

In a study in the mid-1980s from the Thai–Burmese border, the failure rate in school children taking doxycycline prophylaxis was less than for chloroquine (0.8% per week compared to 6.4% per week) (12). Doxycycline prophylaxis to Thai soldiers on the Cambodian border was more effective against both *falciparum* and *vivax* malaria compared to combinations of either proguanil/dapsone or pyrimethamine/dapsone (13). The Australian military experience has confirmed that doxycycline is also effective for preventing *falciparum* malaria on Papua New Guinea (14).

Indications

Tetracyclines are primarily used as a supplement to quinine in the treatment of *P. falciparum* malaria in areas with decreased susceptibility for quinine (i.e. in Thailand or adjacent countries).

Doxycycline is also used for prophylaxis against *P. falciparum* in areas where mefloquine resistance is frequent.

Side effects

Mild gastrointestinal symptoms such as epigastric pain, nausea, vomiting, diarrhoea, and pruritus ani are frequent. Irritative diarrhoea due to the substance should be distinguished from pseudomembraneous colitis due to the overgrowth of *Clostridium difficile* (2). Pronounced photosensitivity reactions of the skin occur and seem to be more common with more intense sun exposure (15). Renal function, in patients with renal damage, is worsened by tetracycline (not by doxycycline). Children under 8 years may develop permanent brown discoloration of the teeth. Liver damage can also be seen especially in pregnant women.

Rare hypersensitivity skin reactions such as urticaria, angio-edema and serum sickness may occur (2).

Contraindications and precautions

Tetracycline should not be given to subjects with pre-existing kidney or liver damage. Doxycycline, however, can be given in unchanged dose to patients with severe kidney failure. Due to tooth staining, tetracyclines should not be given to children under 8 years unless absolutely necessary. Tetracyclines also deposit in the human skeleton and may cause growth retardation, but this effect is reversible in short-term treatment.

Pregnancy and lactation

Tetracyclines are in principle contraindicated during pregnancy, as they may interfere with the teeth and bone development of the foetus (16). However, they may be used if there is a strong therapeutic indication for use.

Tetracyclines are readily excreted into breast milk, but absorption by the infant is probably negligible due to binding to calcium in the milk (6).

Interactions

Iron, milk, and antacids reduce the bioavailability of tetracyclines. The concurrent use of tetracycline with methoxyflurane is nephrotoxic and should be avoided (17). There are no negative reports concerning the combination of tetracyclines with other antimalarials.

Dosage

Prophylaxis

The prophylactic dose of doxycycline is 100 mg once daily to adults and children above 12 years of age. If doxycycline is given to children between 8 and 12 years old the dose is 1.5 mg/kg. Prophylaxis should be started the day before travel and continue for 4 weeks after leaving the area (18). Tetracycline is not used for prophylaxis.

Treatment

Tetracyclines should always be combined with quinine or possibly mefloquine in the treatment of multiresistant *P. falciparum* malaria.

Tetracycline

Adults and children >8 years
250 mg 4 times daily for 7 days.

Doxycycline

Adults and children >8 years
200 mg daily for 7–10 days (19).

Preparations

Many tetracycline and doxycycline preparations are available apart from those mentioned below.

Tetracycline

- Achromycin® (Lederle). Tablets and capsules 250 mg, 500 mg.

Doxycycline

- Vibramycin® (Pfizer). Tablets and capsules 100 mg, 200 mg. Oral suspension 10 mg/ml.

References

1. *Practical Chemotherapy of Malaria*. Report of a WHO Scientific group.Technical Report Series no. 805 (1990). (Geneva: World Health Organization), pp. 35–37.
2. Sande MA, Mandell GL (1990). Antimicrobial agents. In: *Goodman & Gilman's The Pharmacological Basis of Therapeutics*, 8th edn, edited by A. G. Gilman, T. W. Rall, A. S. Nies and P. Taylor, (New York: Pergamon Press), pp. 1117–1124.
3. Hermansson J (1982). Rapid determination of tetracycline and lumecycline in human plasma and urine using high-performance liquid chromatography. *J Chromatogr*, **232**, 385–393.
4. De Leenheer AP, Nelis HJ (1979). Doxycycline determination in human serum and urine by high-performance liquid chromatography. *J Pharm Sci*, **68**, 999–1002.
5. Sheridan ME, Clarke GS (1988). Improved high-performance liquid chromatographic determination of doxycycline in serum and urine using solid-phase extraction columns. *J Chromatogr*, **434**, 253–258.
6. Tetracycline hydrochloride. *Therapeutic Drugs*, edited by Sir Colin Dollery (1991), (London: Churchill Livingstone), pp. T28–T32.
7. Saivin S, Houin G (1988). Clinical pharmacokinetics of doxycycline and minocycline. *Clin Pharmacokinet*, **15**, 355–366.
8. Colwell EJ, Hickman RL, Intraprasert R, Tirabutana C (1972). Minocycline and tetracycline treatment of acute *falciparum* malaria in Thailand. *Am J Trop Med Hyg*, **21**, 144–149.
9. Reacher M, Campbell CC, Freeman J, Doberstyn EB, Brandling-Bennett AD (1981). Drug therapy for *P. falciparum* malaria resistant to pyrimethamine-sulphadoxine (Fansidar): a study of alternate regimens in Eastern Thailand. *Lancet*, **2**, 1066–1069.
10. Pinichpongse S, Doberstyn EB, Cullen JR, Yisunsri L, Thongsombun Y, Thimasarn K (1982). An evaluation of five regimens for the outpatient therapy of *falciparum* malaria in Thailand 1980–81. *Bull World Health Organ*, **60**, 907–912.
11. Looareesuwan S, Viravan C, Vanijanonta S, Wilairatana P, Charoenlarp P, Canfield CJ, Kyle DE (1994). Randomized trial of mefloquine-doxycycline and artesunate-doxycycline for treatment of acute uncomplicated *falciparum* malaria. *Am J Trop Med Hyg*, **50**, 784–789.
12. Pang LW, Boudreau EF, Limsomwong N, Singharaj P (1987). Doxycycline prophylaxis for *falciparum* malaria. *Lancet*, **23**, 1161–1164.
13. Shanks GD, Edstein MD, Suriyamongkol V, Timsad S, Webster HK (1992). Malaria chemoprophylaxis using proguanil/dapsone combinations on the Thai–Cambodian border. *Am J Trop Med Hyg*, **46**, 643–648.
14. Rieckmann KH, Yeo AET, Davis DFR, Hutton DC, Wheatley PF, Simpson R (1993). Recent military experience with malaria chemoprophylaxis. *Med J Aust*, **158**, 446–449.
15. *Meyler's Side Effects of Drugs*, 12th edn (1993), edited by M. N. Dukes (Amsterdam: Elsevier), pp. 160–161, 212–216.
16. Cohlan SQ (1977). Tetracyclines: staining of teeth. *Teratology*, **15**, 127–131.
17. Kuzneu EY (1970). Methoxyflurane, tetracycline and renal failure. *J Am Med Ass*, **221**, 62–64.
18. *International Travel and Health: vaccination requirements and health advice* (1994). (Genveva: World Health Organization).
19. Anonymous (1990). Severe and complicated malaria. *Trans R Soc Trop Med Hyg*, **84**. 1–65.

Thiabendazole

Chemical structure

Thiabendazole

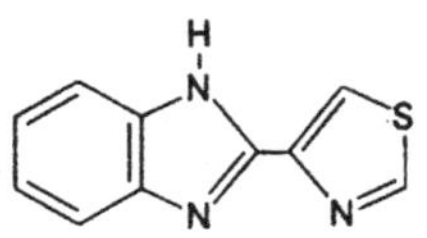

Physical properties

MW 201, pKa not known. Practically insoluble in water.

Pharmacology and mechanism of action

Thiabendazole is a benzimidazole derivative introduced as a veterinary drug during the 1960s and later as a human anthelminthic drug. It has a broad spectrum anthelminthic activity being effective against various types of nematode infections. It is both ovicidal and larvicidal. It is also highly effective against many saprophytic and pathogenic fungi *in vitro* and has also shown anti-inflammatory, antipyretic and analgesic properties in laboratory animals (1). Clinically, it is primarily used against *Strongyloides stercoralis* and cutaneous larva migrans.

The mechanism of action is not clearly understood. It has been shown to inhibit the mitochondrial fumurate reductase, which is specific for helminths (2). Thiabendazole may also affect parasite microtubules, by a mechanism similar to that described for mebendazole (see Mebendazole).

Pharmacokinetics

A specific HPLC method has been described for the determination of thiabendazole and its metabolites, but no pharmacokinetic data based on this method have been published (3).

The drug can be administered orally, topically and rectally. Using unspecific assays the disposition of the drug was studied in four healthy volunteers given 1.0 g of ^{14}C-labelled thiabendazole. The drug was quickly absorbed and peak plasma levels of 13–18 μg/ml were obtained 1–2 hours after drug administration. Drug plasma levels declined rapidly and approached essentially zero between 24 and 48 hours. About 87% of the radioactivity was excreted with the urine during the first 48 hours mainly in the form of glucuronides or sulphates of 5-hydroxythiabendazole. Less than 1% was excreted as the parent drug. A further 5% of the radioactivity was excreted with the faeces during the same period (4). Thiabendazole has also been reported to be significantly absorbed and drug levels sustained for longer periods of time after rectal administration attained peak plasma levels after 4 hours (5). In an anephric patient with *Strongyloides* infection, the pharmacokinetics of thiabendazole and its metabolites were determined during haemodialysis and haemoperfusion after giving a single oral dose of the drug. The plasma half-life, volume of distribution, and clearance for thiabendazole were 1 hour, 2.8 l/kg, and 27 ml/min/kg, respectively. Haemodialysis and haemoperfusion removed efficiently thiabendazole and the 5-hydroxythiabendazole, but were unable to remove the glucuronide and sulphate conjugates (6).

Clinical trials

In an open clinical study, 88 patients with *Strongyloides stercoralis* who had been prisoners of war in the Far East had been treated with thiabendazole 25 mg/kg twice daily for 3 days. The drug was generally reported to be effective (7). In a randomized trial, thiabendazole (50 mg/kg given twice daily for 3 days) was compared to ivermectin (200 μg/kg given as a single dose in 1 or 2 days). Three months after treatment, only 1 of 34 subjects who received thiabendazole, and 2 of 19 patients who received ivermectin had stool positives for *Strongyloides* larvae (8). Against *Trichostrongylus* species, a single dose of thiabendazole, 50 mg/kg had a cure rate above 90% (9). Similar results have been reported after treatment of creeping eruptions (cutaneous larva migrans) with topical applications of 14% thiabendazole suspensions applied 5–6 times for 2 weeks (10) as well as oral treatment with thiabendazole, 25 mg/kg twice daily for 3 days (11). Amelioration of the disease has been reported during treatment of *Capillaria philippinensis* (12) and *Trichinella spiralis* (13).

The results of thiabendazole treatment of *Ascaris lumbricoides* (9, 14), *Ancylostoma duodenale* (9), and *Necator americanus* (14) infections have been variable. Questionable results have also been reported in patients with *Dracunculus medinensis* (15) and *Trichuriasis* (9, 16).

Indications

Thiabendazole is primarily indicated in infections with *Strongyloides stercoralis* and cutaneous larva migrans. It may also prove useful against *Capillaria philippinensis*, *Trichostrongylus* species and alleviate symptoms during the invasion stage of trichinosis.

Pregnancy and lactation

Teratogenicity has not been reported in laboratory animals such as mice, rats, rabbits, swine, sheep and cattle (17, 18, 19). However, one study has reported the drug to be teratogenic in mice after giving high doses during early pregnancy (20). No ill effects have been reported in a baby born to a mother treated with the drug during pregnancy (17). Because of its toxicity and reported teratogenicity in mice, the drug should be avoided during pregnancy, unless there is a strong indication for use.

Its excretion into breast milk is unknown.

Interactions

In a single patient, thiabendazole has been reported to have increased the plasma half-life of theophylline by three-fold because of decreased plasma clearance (21).

Side effects

Common side effects include nausea, vomiting, headache, dizziness, and abdominal pain. In one clinical study around 40% of the patients treated experienced side effects including vomiting (25%), headache (11%) and dizziness (11%) (9). In another study (22), 43 patients treated with the recommended doses of the drug, 34 (89%) suffered side effects. Major complaints included nausea (67%), smelly urine (26%), neuropsychiatric symptoms (23%), malaise (16%), dizziness (16%), anorexia (7%), vomiting (7%), abdominal pain (7%), 'thought going to die' (7%), and headache (5%). The patients in the study were largely elderly. Side effects occurred 1–4 hours after drug ingestion and lasted for up to 8–12 hours.

Occasionally cholestatic jaundice, skin reactions, crystalluria, diarrhoea, headache, fatigue, drowsiness and drying of mucous membranes may occur. Hyperglycaemia, disturbances in colour vision, bradycardia and hypotension are uncommon. Hypersensitivity reactions such as fever, oedema, and lymphoadenopathy are also rare (9, 23). Single cases of Stevens–Johnsons syndrome and toxic epidermal necrosis have been reported (23). The urine of some patients may have an odour much like that observed after eating asparagus; it is attributed to the presence of a metabolite (1, 17).

Contraindications and precautions

Thiabendazole should be given with caution to patients with a history of drug hypersensitivity. Dosage reductions must be made in patients with kidney or hepatic failure. Thiabendazole is a potent inhibitor of cytochrome P450, and it is likely to increase the plasma concentrations of drugs metabolized by this route.

Dosage

Strongyloides stercoralis, Trichostrongylus spp., Trichinella spiralis

25 mg/kg twice daily for 3 days (maximum daily dose is 3 g). In disseminated strongyloidiasis the treatment can be extended to 5 days.

Capillaria philippinensis

25 mg/kg once daily for 30 days.

Cutaneous larva migrans

10% suspension of thiabendazole is applied on affected skin areas 5–6 times/day until regression occurs. Thereafter 3 times daily over 2 weeks. Otherwise, 25 mg/kg twice daily for 3 days (maximum daily dose 3 g) orally.

Preparations

- Mintesol® (Merck Sharp & Dohme). Oral suspension 100 mg/ml. Tablets 500 mg.

References

1. Robinson HJ, Phases HF, Graessle DE (1969). Thiabendazole: toxicological, pharmacological and antifungal properties. *Texas Rep Biol Med*, **27**, 537–560.
2. Sheth UK (1975). Thiabendazole inhibited the fumarate reductase metabolism of helminths. *Prog Drug Res*, **19**, 147.
3. Watts MT, Raisys VA, Bauer LA (1982). Determination of thiabendazole and 5-hydroxythiabendazole in human serum by fluorescence-detected high-performance liquid chromatography. *J Chromatogr*, **230**, 79–86.
4. Tocco DJ, Rosenblum C, Martin CM, Robinson HJ (1966). Absorption, metabolism, and excretion of thiabendazole in man and laboratory animals. *Toxicol Appl Pharmacol*, **9**, 31–39.

5. Boken DJ, Leoni PA, Preheim LC (1993). Treatment of *Strongyloides stercoralis* hyperinfection syndrome with thiabendazole per rectum. *Clin Infect Dis*, **16**, 123–126.
6. Bauer LA, Raisys VD, Watts MT, Ballinger J (1982). The pharmacokinetics of thiabendazole and its metabolites in anephric patient undergoing hemodialysis and hemoperfusion. *J Clin Pharmacol*, **22**, 276–280.
7. Gill GV, Bell DR (1979). *Strongyloides stercoralis* infection in former Far East prisoners of war. *BMJ*, **2**, 572–574
8. Gann PH, Neva FA, Gam AA (1994). A randomized trial of single- and two-dose ivermectin versus thiabendazole for treatment of strongyloidiasis. *J Infect Dis*, **169**, 1076–1079.
9. Farahmandian I, Arfaa F, Jalali H, Reza M (1977). Comparative studies on the evaluation of the effect of new anthelminthics on various intestinal helminthiasis in Iran. *Chemotherapy*, **23**, 98.
10. Whiting DA (1976). The successful treatment of creeping eruptions with topical thiabendazole. *S Afr Med J*, **50**, 253–255.
12. Whalen GE, Strickland GT, Cross JH, Rosenberg EB, Gutman RA, Watten RH, Uylanggco C, Dizon JJ (1969). Intestinal capillariasis. A new disease in man. *Lancet*, **i**, 13–16.
13. Clark PS, Brownsberger KM, Saslow AR, Kagan IG, Noble GR, Maynard JE (1972). Bear meat trichinosis. Epidemiologic, serologic, and clinical observations from two Alaskan outbreaks. *Ann Intern Med*, **76**, 951–956.
14. Kale OO (1977). A comparative trial of the anthelminthic efficacy of pyrantel pamoate (combantrin) and thiabendazole (Mintezol), *Afr J Med Sci*, **6**, 89–93.
15. Belcher DW, Wurapa FK, Ward WB (1975). Failure of thiabendazole and metronidazole in the treatment and suppression of guinea worm disease. *Am J Trop Med Hyg*, **24**, 444–446.
16. Stuart JE, Welch JS (1973). Trial of thiabendazole and viprynium embonate in combination for trichuriasis in aboriginal children. *Med J Aust*, **2**, 1017–1019.
17. Robinson HJ, Phases HF, Graessle DE (1978). The toxicological and antifungal properties of thiabendazole. *Ecotoxicol Environ Safety*, **1**, 471–476.
18. Wise LD, Cartwright ME, Seider CL, Sachuk LA, Lankas GR (1994). Dietary two-generation reproduction study of thiabendazole in Sprague-Dawley rats. *Food Chem Toxicol*, **32**, 239–246.
19. Lankas GR, Wise DL (1993). Developmental toxicity of orally administered thiabendazole in Sprague–Dawley rats and New Zealand white rabbits. *Food Chem Toxicol*, **31**, 199–207.
20. Ogata A, Ando H, Kubo Y, Hiraga K (1984). Teratogenicity of thiabendazole in ICR mice. *Food Chem Toxicol*, **22**, 509–520.
21. Schneider D, GannonR, Sweeney K, Shore E (1990). Theophylline and antiparasitic drug interactions. A case report and study of the influence of thiabendazole and mebendazole on theophylline pharmacokinetics in adults. *Chest*, **97**, 84–87.
22. Grove DI (1982). Treatment of strongyloidiasis with thiabendazole: an analysis of toxicity and effectiveness. *Trans R Soc Trop Med Hyg*, **76**, 114–118.
23. Robinson HM, Samorodin CS (1976). Thiabendazole-induced toxic epidermal necrolysis. *Arch Dermatol*, **112**, 1757–1760.

Tinidazole

Chemical structure

Tinidazole

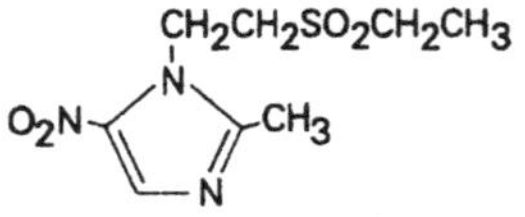

Physical properties

MW 247; pKa 1.8. The drug is practically insoluble in water.

Pharmacology and mechanism of action

Similar to metronidazole.

Pharmacokinetics

Specific GC (1) and HPLC (2) methods have been described for the determination of tinidazole and its metabolites.

Tinidazole is given orally, parenterally and as suppository. Its oral bioavailability is close to 100% (3). The suppositories have a bioavailability of 60–70% (4). Peak plasma concentrations of around 40 μg/ml are achieved within 2 hours of a 2 g dose. Tinidazole is widely distributed and concentrations similar to those in plasma have been achieved in bile, breast milk, cerebrospinal fluid and saliva. It crosses the placenta readily. It has an apparent volume of distribution of around 0.6 l/kg, and is only slightly bound to plasma proteins (<20%) (5).

The plasma elimination half-life is around 13 hours (5, 6). After an intravenous infusion of 800 mg tinidazole to 2 healthy volunteers, 40% of the radioactivity was excreted with the urine during the first 24 hours, which increased to 63% after 5 days. About 12% of the radioactivity was excreted with the faeces indicating a biliary excretion of the drug. Unchanged tinidazole accounted for a mean of 32% of the total radioactivity excreted with the urine. Ethyl 2-(5-hydroxy-2-methyl-4-nitro-1-imidazolyl)ethyl sulphone was the major metabolite in urine (about 30% of radioactivity). 2-hydroxytinidazole was excreted as a minor metabolite (about 9% of the radioactivity). Unchanged tinidazole and the major urinary metabolite were also present in the faeces (7).

Tinidazole is effectively removed by haemodialysis and thus dose supplements are necessary after dialysis (8, 9).

Clinical trials

The cure rate in acute amoebiasis, amoebic liver abscess and in giardiasis is close to 100% at the dose recommended below (10–15). In an open study, Simjee *et al.* (16) compared the efficacy of metronidazole *vs* tinidazole. Of 75 patients with amoebic liver abscess, 48 patients received metronidazole and 27 were given tinidazole. A single daily dose of 2 g for 5 days was given for both drugs. Both drugs were highly effective and resulted in rapid clinical improvement. However, 4 patients treated with tinidazole and 2 patients treated with metronidazole required a second course of therapy. In another study, Speelman (17) compared the efficacy of metronidazole against tinidazole in giardiasis. Of 63 patients treated, 31 received tinidazole (single dose of 50 mg/kg to a maximum of 2 g) and 32 received metronidazole (single dose of 60 mg/kg to a maximum of 2.4 g or 50 mg/kg to a maximum of 2 g for 3 days). In both groups, tinidazole was more effective with a cure rate of above 93%.

Indications

Infections caused by *Entamoeba histolytica* and *Giardia lamblia*. Tinidazole is more effective than metronidazole in the treatment of giardiasis.

Pregnancy and lactation

Teratogenicity has not been reported in rats and mice (18). Documentation in man is lacking. The drug passes the placenta barrier and reaches the fetus attaining concentrations half of those in plasma (19). Since tinidazole has similar pharmacological properties to metronidazole it should be avoided during the first trimester, unless there is a strong therapeutic indication.

Tinidazole is excreted into breast milk (5). During lactation, nursing should be suspended temporarily.

Side effects

Side effects are similar to but milder than those caused by metronidazole. Gastrointestinal disturbances like nausea, vomiting, anorexia and metallic taste are common. Headache, tiredness, furred tongue and itching may occur. Thrombophlebitis may occur at the site of intravenous infusion (20).

Contraindications

Tinidazole should not be taken together with alcohol.

Interactions

A disulfiram-like reaction might occur if tinidazole is taken together with alcohol.

Dosage

Giardiasis:

Adults
2 g as a single oral dose.
Children
50 mg/kg as a single oral dose.

Amoebiasis

Adults
2 g as single daily dose orally for 5 days.
Children
50–75 mg/kg as a single daily dose orally for 5 days.

In invasive amoebiasis, following treatment with tinidazole, diloxanide 500 mg 3 times daily for 10 days must be given to eradicate residual amoeba in the lumen.

Preparations

- Fasigyn® (Pfizer). Tablets 150 mg, 200 mg, 300 mg, 500 mg, 1 g. Oral suspension 200 mg per ml. Solution for injection 2 mg per ml.
- Tricolam® (Pfizer). Tablets 500 mg.
- Simplotan® (Pfizer). Tablets 1 g.

References

1. Bhatia SC, Shanbhag VD (1984). Electron-capture gas chromatographic assays of 5-nitroimidazole class of antimicrobials in blood. *J Chromatogr*, **305**, 325–334.
2. Gibson RA, Lattaziol L, McGee H (1984). Optimized liquid-chromatographic determination of metronidazole and its metabolism in plasma. *Clin Chem*, **30**, 784–787.
3. Vinge E, Andersson KE, Ando G, Lunell E (1983). Biological availability and pharmacokinetics of tinidazole after single and repeated doses. *Scand J Infect Dis*, **15**, 391–397.
4. Mattila J, Männistö PT, Mäntylä R, Nykänen S, Lamminsivu U (1983). Comparative pharmacokinetics of metronidazole and tinidazole as influenced by administration route. *Antimicrob Agents Chemother*, **23**, 721–725.
5. Wood BA, Faulkner JK, Monro AM (1982). The pharmacokinetics, metabolism and tissue distribution of tinidazole. *J Antimicrob Chemother*, **10**, 43–57.
6. Klimowicz A, Nowak A, Bielecka-Grzela S (1992). Penetration of tinidazole into skin blister fluid following its oral administration. *Eur J Clin Pharmacol*, **43**, 523–526.
7. Wood S.G, John BA, Chasseaud LF, Brodie RR, Baker JM, Faulkner JK, Wood BA, Darragh A, Lambe RF (1985). Pharmacokinetics and metabolism of ^{14}C-tinidazole in humans. *J Antimicrob Chemother*, **17**, 801–809.
8. Flouvat BL, Imbert C, Dubois DM, Temperville BP, Roux AF, Chevalier GC, Humbert C (1983). Pharmacokinetics of tinidazole in chronic renal failure and in patients on haemodialysis. *J Clin Pharmacol*, **15**, 735–741.
9. Robson RA, Bailey RR, Sharman JR (1984). Tinidazole pharmacokinetics in severe renal failure. *Clin Pharmacokinet*, **9**, 88–94.
10. Gazder AT, Banerjee M (1978). Single dose therapy of giardiasis with tinidazole and metronidazole. *Drugs*, **15**, 30–32.
11. Islam N, Hassan K (1978). Tinidazole and metronidazole in hepatic amoebiasis. *Drugs*, **15**, 26–29.
12. Joshi HD, Shah BM (1975). A comparative study of tinidazole and metronidazole in treatment of amoebiasis. *Indian Practitioner*, **28**, 295.
13. Khokhani RC, Garud AD, Deodhar KP, Sureka SB, Kulkarni M, DamLe VB (1978). Treatment of amoebic liver abscess with tinidazole and metronidazole. *Drugs*, **15**, 23–25.
14. Leerman ST, Walker RA (1982). Treatment of Giardiasis – literature review and recommendations. *Clin Paed*, **21**, 409–414.
15. Misura NP, Laiq SM (1974). Comparative trial of tinidazole and metronidazole in intestinal amoebiasis. *Curr Ther Res Clin Exp*, **16**, 1255–1263.
16. Simjee AE, Gathiram V, Jackson TFHG, Khan BFY (1985). A comparative trial of metronidazole v. tinidazole in the treatment of amoebic abscess. *S Afr Med J*, **68**, 923–924.

17. Speelman P (1985). Single-dose tinidazole for the treatment of giardiasis. *Antimicrob Agents Chemother*, **27**, 227–229.
18. Owaki Y, Momiyama H, Sakai T, Nabato II (1974). Effects of tinidazole on the fetuses and their post-natal development in mice and rats. *Pharmacometrics*, **8**, 421.
19. Karhunen M (1984). Placental transfer of metronidazole and tinidazole in early human pregnancy after a single infusion. *Br J Clin Pharmacol*, **18**, 254–257.
20. Sawyer PR, Brogden RN, Pinder RM, Speight TM, Avery GS (1976). Tinidazole: a review of its antiprotozoal activity and therapeutic efficacy. *Drugs*, **11**, 424–440.

Index